AF824799

CliffsNotes® Algebra I Common Core Quick Review

CliffsNotes® Algebra I Common Core Quick Review

By Kimberly Gores, M.A.T.

About the Author
Kimberly Gores, M.A.T., has been a mathematics educator and researcher for over 15 years. Gores has taught introductory and advanced algebra and mathematics classes at the middle school, high school, and college levels. She is currently a math teacher at Coeur d'Alene High School, Idaho.

Editorial
Executive Editor: Greg Tubach
Senior Editor: Christina Stambaugh
Production Editor: Erika West
Copy Editor: Lynn Northrup
Technical Editors: Mary Jane Sterling and Tom Page
Proofreader: Pamela Weber-Leaf
Indexer: Potomac Indexing, LLC

CliffsNotes® Algebra I Common Core Quick Review

Copyright © 2022 by Learneo, Inc.

All rights reserved.

Library of Congress Control Number: 2016943153
ISBN: 978-1-957671-13-0

Printed in the United States of America
DOC 10 9 8 7 6 5 4 3 2 1

Note: If you purchased this book without a cover, you should be aware that this book is stolen property. It was reported as "unsold and destroyed" to the publisher, and neither the author nor the publisher has received any payment for this "stripped book."

The publisher and the author make no representations or warranties with respect to the accuracy or completeness of the contents of this work and specifically disclaim all warranties, including without limitation warranties of fitness for a particular purpose. No warranty may be created or extended by sales or promotional materials. The advice and strategies contained herein may not be suitable for every situation. This work is sold with the understanding that the publisher is not engaged in rendering legal, accounting, or other professional services. If professional assistance is required, the services of a competent professional person should be sought. Neither the publisher nor the author shall be liable for damages arising here from. The fact that an organization or Website is referred to in this work as a citation and/or a potential source of further information does not mean that the author or the publisher endorses the information the organization or Website may provide or recommendations it may make. Further, readers should be aware that Internet Websites listed in this work may have changed or disappeared between when this work was written and when it is read.

Trademarks: CliffsNotes, the CliffsNotes logo, Cliffs, CliffsAP, CliffsComplete, CliffsQuickReview, CliffsStudySolver, CliffsTestPrep, CliffsNote-a-Day, cliffsnotes.com, and all related trademarks, logos, and trade dress are trademarks or registered trademarks of Learneo, Inc., and/or its affiliates. All other trademarks are the property of their respective owners. Learneo, Inc. is not associated with any product or vendor mentioned in this book..

Common Core State Standards © Copyright 2010. National Governors Association Center for Best Practices and Council of Chief State School Officers. All rights reserved.

This product is not sponsored or endorsed by the Common Core State Standards Initiative of the National Governors Association Center for Best Practices and the Council of Chief State School Officers.

Table of Contents

Chapter 1

FOUNDATIONS FOR UNDERSTANDING ALGEBRA

Chapter Check-In

- ❑ Real numbers and the number system
- ❑ Properties of operations
- ❑ Powers and exponents
- ❑ Square roots and cube roots
- ❑ Evaluating numerical expressions with grouping symbols and following the order of operations
- ❑ Divisibility rules
- ❑ Positive and negative signed numbers
- ❑ Absolute value
- ❑ Fractions
- ❑ Decimals
- ❑ Percents and proportions
- ❑ Scientific notation

Algebra can be defined as arithmetic operations using variables and/or symbols in place of numbers. Algebra is built upon your earlier understandings of arithmetic. As you get closer to entering the world of algebra, and before you begin learning or relearning algebra, you should have a solid background in pre-algebra terms, properties, and operations. Topics covered in this chapter will provide you with the foundational concepts necessary to increase your algebraic proficiency and make sense of multi-step Common Core Mathematics problems discussed in Chapter 12, "Application Problems."

Foundations for Understanding Numbers

The first items you should become familiar with are the different types of numbers and the common math symbols.

Common Core Standard: The Real Number System

Use properties of rational and irrational numbers to explain why (a) the sum or product of two rational numbers is rational; (b) the sum of a rational number and an irrational number is irrational; and (c) the product of a nonzero rational number and an irrational number is irrational (N.RN.3).

Real numbers and the number system

In expressing Common Core Mathematic operations, you work with several types of rational and irrational numbers. **Real numbers,** or all points on a number line, can be grouped or represented in many different ways. Each of these groups has its own set of characteristics and rules to express how these numbers relate to other numbers.

- **Natural** or **counting numbers.** The numbers 1, 2, 3, 4, ... are called natural or counting numbers. ***Note:*** Zero is *not* a natural number.
- **Whole numbers.** The numbers 0, 1, 2, 3, ... are called whole numbers.
- **Integers.** The numbers ... −2, −1, 0, 1, 2, ... are called integers. All integers are rational numbers and can be negative or positive. ***Note:*** Zero is neither positive nor negative. The terms *even* and *odd* apply only to integers.

  Even numbers. Even numbers are integers that are divisible by 2: ... −6, −4, −2, 0, 2, 4, 6, ...

  Odd numbers. Odd numbers are integers that are *not* divisible by 2: ... −5, −3, −1, 0, 1, 3, 5, ...
- **Rational numbers.** Fractions such as $\frac{3}{2}$ or $\frac{7}{8}$ are called rational numbers. Since a number such as 5 may be written as $\frac{5}{1}$, all integers are rational numbers. However, not all rational numbers are integers, because only rational numbers can be fractions. All rational numbers can be written as a ratio of two integers, $\frac{a}{b}$, where $b \neq 0$. Terminating and repeating decimals are also rational numbers, because they can be written as fractions in this form.

- **Irrational numbers.** Another type of number is an irrational number. In Common Core Mathematics, you should be able to use rational numbers to approximate decimal values of irrational numbers. Irrational numbers *cannot* be written as a ratio of two integers, $\frac{a}{b}$, where $b \neq 0$. Both $\sqrt{3}$ and π are examples of irrational numbers. An irrational number, when expressed as a decimal, neither terminates nor has a repeating decimal pattern.
- **Prime numbers.** A prime number is a natural number greater than 1 that has exactly two factors, or that can be evenly divided by only itself and 1. For example, 19 is a prime number because it can be evenly divided by only 19 and 1, but 21 is not a prime number because 21 can be evenly divided by other numbers (3 and 7). The only even prime number is 2; thereafter, any even number may be evenly divided by 2. Zero and 1 are neither prime numbers nor composite numbers. The first 10 prime numbers are 2, 3, 5, 7, 11, 13, 17, 19, 23, and 29.
- **Composite numbers.** A composite number is a natural number divisible by more than just 1 and itself: … 4, 6, 8, 9, …
- **Square numbers (perfect squares).** Square numbers or perfect squares are nonzero integers that result when numbers are multiplied by themselves; that is, raised to the second power (squaring them). For example, $2^2 = 2 \cdot 2 = 4$; $3^2 = 3 \cdot 3 = 9$. The first six squares of positive, natural numbers are 1, 4, 9, 16, 25, and 36. ***Note:*** All numbers that are squares have equal positive factors and equal negative factors. For example, $(-1)^2 = (-1)(-1) = 1$ and $(1)^2 = (1)(1) = 1$. (***Note:*** Multiplying two negative numbers always gives a positive answer.)
- **Cube numbers (perfect cubes).** Cube numbers are the result when numbers are multiplied by themselves and then again by the original number; that is, raised to the third power (cubing them). $2^3 = 2 \cdot 2 \cdot 2 = 8$; $3^3 = 3 \cdot 3 \cdot 3 = 27$. The first six cubes of natural numbers are 1, 8, 27, 64, 125, and 216.

Ways to show multiplication expressions

There are several ways to interpret and see the multiplication structure of a pair of numerical **values.**

- When the two numerical values are known, such as 4 and 3, you can show the multiplication statement as follows:

$$4 \times 3$$
$$4 \cdot 3$$
$$(4)(3)$$
$$4(3)$$
$$(4)3$$

- When one value is a number and the other value is a **variable,** such as 4 and *a,* you can show the expression of the multiplication statement as follows:

$$4 \times a$$
$$4 \cdot a$$
$$(4)(a)$$
$$4(a)$$
$$(4)a$$
$$4a$$

- When both values are variables, such as *a* and *b,* you can show the expression of the multiplication statement as follows:

$$a \times b$$
$$a \cdot b$$
$$(a)(b)$$
$$a(b)$$
$$(a)b$$
$$ab$$

Common math symbols

The following math symbols appear throughout Common Core Mathematics algebra. Be sure to know what each symbol represents.

Symbol	*Written Expression*
$=$	is equal to
$\neq$	is not equal to
$\approx$ or $\doteq$	is approximately equal to
$>$	is greater than
$\geq$ or $\geqq$	is greater than or equal to
$\ngtr$	is not greater than
$\ngeq$	is not greater than or equal to
$<$	is less than
$\leq$ or $\leqq$	is less than or equal to
$\nless$	is not less than
$\nleq$	is not less than or equal to

Properties of Basic Mathematical Operations and Algebraic Thinking

Some mathematical operations apply previous understandings of arithmetic properties that can make algebraic thinking easier to understand and can actually save you time.

Use the table below as a reference for basic math vocabulary terms that signal a math operation.

Vocabulary	*Definition*
sum	The answer to an addition problem.
difference	The answer to a subtraction problem.
product	The answer to a multiplication problem.
quotient	The answer to a division problem.
dividend	A number or quantity being divided.
divisor	A number by which the dividend is divided.
addend	A number that is added in an addition problem.
subtrahend	A number being subtracted from another number.
minuend	A number from which the subtrahend is to be subtracted.
factor	One of the numbers multiplied in a multiplication problem.
term	A single number or variable, or numbers and variables, multiplied together.
divisible	When one number can be divided by another with no remainder.

Some properties of addition

You should know the definition of each of the following properties of addition and how each can be applied.

- **Closure** is when results of an operation performed on a set of numbers fall into the original set. If you add two even numbers, the answer is still an even number ($2 + 4 = 6$); therefore, the set of even numbers is *closed* under addition (has closure). If you add two odd numbers, the answer is not an odd number ($3 + 5 = 8$); therefore, the set of odd numbers is *not closed* under addition (no closure).

- **Commutative** means that the order of the numbers does not make any difference in the result when adding.

$$2+3=3+2$$
$$a+b=b+a$$

Note: The commutative property does not hold for subtraction because for any two real numbers, a and b, there are cases where $a - b \neq b - a$. For example:

$$3-1\neq 1-3$$
$$2\neq -2$$
$$a-b\neq b-a$$

- **Associative** means that the sum stays the same when the grouping of addends is changed.

$$(2+3)+4=2+(3+4)$$
$$(a+b)+c=a+(b+c)$$

The grouping has changed (parentheses moved), but the sides are still equal.

Note: The associative property does not hold for subtraction because for any three real numbers, a, b, and c, there are cases where $a - (b - c) \neq (a - b) - c$. For example:

$$4-(3-1)\neq(4-3)-1$$
$$4-2\neq 1-1$$
$$2\neq 0$$
$$a-(b-c)\neq(a-b)-c$$

- The **identity element for addition,** also known as additive identity, is 0. The sum of any number and 0 gives the original number.

$$3+0=0+3=3$$
$$a+0=0+a=a$$

- The **additive inverse** is the opposite (negative) of the number. A number and its opposite have a sum of 0, meaning they are additive inverses.

$$3+(-3)=0;\text{ therefore, 3 and }-3\text{ are additive inverses.}$$
$$(-2)+2=0;\text{ therefore, }-2\text{ and 2 are additive inverses.}$$
$$a+(-a)=0;\text{ therefore, }a\text{ and }-a\text{ are additive inverses.}$$

Some properties of multiplication

You should know the definition of each of the following properties of multiplication and how each can be applied to generate equivalent expressions.

- **Closure** is when all answers fall into the original set. If you multiply two even numbers, the answer is still an even number ($2 \times 4 = 8$); therefore, the set of even numbers is *closed* under multiplication (has closure). If you multiply two odd numbers, the answer is an odd number ($3 \times 5 = 15$); therefore, the set of odd numbers is *closed* under multiplication (has closure).

- **Commutative** means the order of the numbers does not make any difference in the result when multiplying.

$$2\times 3=3\times 2$$
$$a\times b=b\times a$$

Note: The commutative property does not hold for division because for any two real numbers, a and b, there are cases where $a \div b \neq b \div a$. For example:

$$2\div 4\neq 4\div 2$$
$$\frac{2}{4}\neq\frac{4}{2}$$
$$\frac{1}{2}\neq 2$$
$$a\div b\neq b\div a$$

- **Associative** means that the product stays the same when the grouping of factors is changed.

$$(2 \times 3) \times 4 = 2 \times (3 \times 4)$$
$$(a \times b) \times c = a \times (b \times c)$$

The grouping has changed (parentheses moved), but the sides are still equal.

Note: The associative property does not hold for division because for any three real numbers, *a*, *b*, and *c*, there are cases where $(a \div b) \div c \neq a \div (b \div c)$. For example:

$$(8 \div 4) \div 2 \neq 8 \div (4 \div 2)$$
$$2 \div 2 \neq 8 \div 2$$
$$1 \neq 4$$
$$(a \div b) \div c \neq a \div (b \div c)$$

- The **identity element for multiplication** is 1, also known as the multiplicative identity. Any number multiplied by 1 gives the original number.

$$3 \times 1 = 3$$
$$a \times 1 = a$$

- The **multiplicative inverse** is the **reciprocal** of the number. Any nonzero number multiplied by its reciprocal equals 1.

$2 \times \frac{1}{2} = 1$; therefore, 2 and $\frac{1}{2}$ are multiplicative inverses.

$a \times \frac{1}{a} = 1$; therefore, *a* and $\frac{1}{a}$ are multiplicative inverses (provided $a \neq 0$).

The distributive property

The **distributive property** is the process of multiplying a factor outside of the parentheses by a sum or difference of two or more terms inside the parentheses. To simplify, multiply the factor by each term inside of the parentheses separately and then add each of the products, if possible. In order to apply the distributive property, it must be multiplication outside the parentheses and either addition or subtraction inside the parentheses.

$$\begin{aligned} 2(3+4) &= 2(3)+2(4) \\ 2(7) &= 6+8 \\ 14 &= 14 \\ a(b+c) &= a(b)+a(c) \end{aligned} \qquad \begin{aligned} 5(12-3) &= 5(12)-5(3) \\ 5(9) &= 60-15 \\ 45 &= 45 \\ a(b-c) &= a(b)-a(c) \end{aligned}$$

Note: You cannot use the distributive property with only one operation.

$$\begin{aligned} 3(4\times5\times6) &\neq 3(4)\times3(5)\times3(6) \\ 3(120) &\neq 12\times15\times18 \\ 360 &\neq 3240 \\ a(bcd) &\neq a(b)\times a(c)\times a(d) \\ a(bcd) &\neq (ab)(ac)(ad) \end{aligned}$$

Multiplying and dividing using zero

In Common Core Mathematics, you should understand that integers can be divided, provided that the divisor is not zero. The product of zero and any number equals zero.

$$\begin{aligned} 0\times5 &= 0 \\ 0\times(-3) &= 0 \\ 8\times9\times3\times(-4)\times0 &= 0 \end{aligned}$$

Likewise, the quotient of zero and any nonzero divisor is zero. (***Note:*** Since division is defined to be the inverse operation of multiplication, the following examples will be justified by using multiplication.)

$$\begin{aligned} 0\div5 = 0 &\text{ since } 5\times0 = 0 \\ 0\div(-6) = 0 &\text{ since } 0\times(-6) = 0 \end{aligned}$$

Important note: Dividing by zero is either "undefined" or "indeterminate," depending on the dividend, and is not permitted.

$\frac{6}{0}$ is undefined. The expression $\frac{6}{0} = x$ requires a value to be found for the unknown quantity in $x \cdot 0 = 6$, but since any number multiplied by 0 is 0, there is no number that satisfies the equation.

$\frac{0}{0}$ is indeterminate. The expression $\frac{0}{0} = x$ requires a value to be found for the unknown quantity in $x \cdot 0 = 0$, but since any number multiplied by 0 is 0, any number can satisfy the equation.

Therefore, $\frac{6}{0}$ has no answer, and $\frac{0}{0}$ does not have a unique answer.

The answer is not zero in either case.

Powers and Exponents

An **exponent** is a number placed above and to the right of a quantity. It expresses the *power* to which the quantity is to be raised or lowered. In 4^3, 3 is the exponent and 4 is called the base. It shows that 4 is to be used as a factor of repeated multiplication three times, $4 \times 4 \times 4$. 4^3 is read as *four to the third* power (or *four cubed,* as discussed in the next section, "Squares and cubes").

$$2^4 = 2 \times 2 \times 2 \times 2 = 16$$

$$3^2 = 3 \times 3 = 9$$

Remember that $x^1 = x$ and $x^0 = 1$ when x is any number (other than 0).

$$2^1 = 2 \qquad 2^0 = 1$$

$$\left(\frac{1}{3}\right)^1 = \frac{1}{3} \qquad \left(\frac{1}{3}\right)^0 = 1$$

$$(-4)^1 = -4 \qquad (-4)^0 = 1$$

If the exponent is negative, such as 3^{-2}, then the base can be dropped under the number 1 in a fraction and the exponent made positive. An alternative method is to take the reciprocal of the base and change the exponent to a positive value.

Example 1: Simplify each numerical expression with an equivalent expression in exponential notation, then evaluate the expression.

	Under 1 Method	Reciprocal Method
(a) 3^{-2}	$3^{-2} = \frac{1}{3^2} = \frac{1}{9}$	$3^{-2} = \left(\frac{1}{3}\right)^2 = \frac{1}{9}$
(b) 2^{-3}	$2^{-3} = \frac{1}{2^3} = \frac{1}{8}$	$2^{-3} = \left(\frac{1}{2}\right)^3 = \frac{1}{8}$
(c) $\left(\frac{2}{3}\right)^{-4}$	$\left(\frac{2}{3}\right)^{-4} = \frac{1}{\left(\frac{2}{3}\right)^4} = \frac{1}{\left(\frac{16}{81}\right)} = \frac{81}{16}$	$\left(\frac{2}{3}\right)^{-4} = \left(\frac{3}{2}\right)^4 = \frac{81}{16}$

Squares and cubes

Common Core Standard: The Real Number System

Extend the properties of exponents to rational exponents (N.RN.1-2).

Two specific types of powers should be noted, **squares** and **cubes.**

To square a number, just multiply it by itself (the exponent would be 2). For example, 6 squared (written 6^2) is 6×6, or 36. 36 is called a perfect square (the square of a whole number). Following is a list of the first twelve perfect squares:

$1^2 = 1$	$5^2 = 25$	$9^2 = 81$
$2^2 = 4$	$6^2 = 36$	$10^2 = 100$
$3^2 = 9$	$7^2 = 49$	$11^2 = 121$
$4^2 = 16$	$8^2 = 64$	$12^2 = 144$

To cube a number, just multiply it by itself twice (the exponent would be 3). For example, 5 cubed (written 5^3) is $5 \times 5 \times 5$, or 125. 125 is called a perfect cube (the cube of a whole number). Following is a list of the first twelve perfect cubes.

$1^3 = 1$	$5^3 = 125$	$9^3 = 729$
$2^3 = 8$	$6^3 = 216$	$10^3 = 1{,}000$
$3^3 = 27$	$7^3 = 343$	$11^3 = 1{,}331$
$4^3 = 64$	$8^3 = 512$	$12^3 = 1{,}728$

Properties of exponents

In Common Core Mathematics, you should know and apply properties of integer exponents to generate equivalent numerical expressions, as illustrated in the examples below. To multiply powers with the same base, simply keep the base number and add the exponents.

Example 2: Simplify each expression with an equivalent expression in exponential notation.

(a) $2^3 \times 2^5 = 2^8$ $\quad (2\times2\times2)\times(2\times2\times2\times2\times2) = 2^8$

(b) $3^2 \times 3^4 = 3^6$ $\quad (3\times3)\times(3\times3\times3\times3) = 3^6$

To divide powers with the same base, simply keep the base number and subtract the second exponent from the first, or the exponent of the denominator from the exponent of the numerator.

Example 3: Simplify each expression with an equivalent expression in exponential notation.

(a) $4^8 \div 4^5 = 4^3$ $\quad 4^8 \div 4^5 = \dfrac{4^8}{4^5} = \dfrac{4\cdot4\cdot4\cdot4\cdot4\cdot4\cdot4\cdot4}{4\cdot4\cdot4\cdot4\cdot4} = 4\cdot4\cdot4 = 4^3$

(b) $\dfrac{9^6}{9^2} = 9^4$ $\quad \dfrac{9^6}{9^2} = \dfrac{9\cdot9\cdot9\cdot9\cdot9\cdot9}{9\cdot9} = 9\cdot9\cdot9\cdot9 = 9^4$

To multiply powers with a different base, you must simplify each number with an exponent first and then perform the operation.

Example 4: Simplify and perform the operation indicated.

(a) $3^2 \times 2^2 = 9\times4 = 36$

(b) $6^2 \div 2^3 = 36\div8 = 4\dfrac{4}{8} = 4\dfrac{1}{2}$

(Some shortcuts are possible.)

To add or subtract powers, whether the base numbers are the same or different, you must simplify each power first and then perform the indicated operation.

Example 5: Simplify and perform the operation indicated.

(a) $3^2 - 2^3 = 9 - 8 = 1$

(b) $4^3 + 3^2 = 64 + 9 = 73$

If a power is raised to another power $(4^2)^3$, simply keep the original base number and multiply the exponents.

Example 6: Simplify each expression with an equivalent expression in exponential notation.

(a) $(4^2)^3 = (4^2)(4^2)(4^2) = 4^6$ or $(4^2)^3 = 4^{(2)(3)} = 4^6$

(b) $(3^3)^2 = (3^3)(3^3) = 3^6$ or $(3^3)^2 = 3^{(3)(2)} = 3^6$

Square Roots and Cube Roots

A *radical expression* is an expression that uses a root, such as a square root, a cube root, or some other root. A radical includes a **radical sign,** $\sqrt{\ }$, which indicates the square root of some number. The value of a higher root, such as a cube root, is written as $\sqrt[3]{\ }$. In a radical expression, $\sqrt[n]{a}$, the number under the radical sign, *a*, is the *radicand* and the degree of the root, *n*, is the *index*.

Other roots are defined similarly and identified by the index given. (In a square root, an index of 2 is understood and usually not written.) A radical expression, $\sqrt[n]{a}$, can also be written as a power with a fractional exponent, $a^{\frac{1}{n}}$. Note that operations using square roots and cube roots are often included in algebra sections, and both topics are discussed more in Chapter 10, "Simplifying Radicals."

Square roots

To find the **square root** of a number, you want to find some number that when multiplied by itself gives you the original number. For example, to find the square root of 25, you want to find the number that when multiplied by itself gives you 25. The square root of 25, then, is 5. Using the proper notation, $\sqrt{25} = \sqrt{5^2} = 5$. Following is a list of the first eleven square roots of perfect squares.

$\sqrt{0} = 0$	$\sqrt{16} = 4$	$\sqrt{64} = 8$
$\sqrt{1} = 1$	$\sqrt{25} = 5$	$\sqrt{81} = 9$
$\sqrt{4} = 2$	$\sqrt{36} = 6$	$\sqrt{100} = 10$
$\sqrt{9} = 3$	$\sqrt{49} = 7$	

Special note: If no sign (or a positive sign) is placed in front of the radical, then the positive answer is required. Only if a negative sign is in front of the radical is the negative answer required. This notation is used in many texts and is adhered to in this book. Therefore,

$$\sqrt{9} = 3 \qquad \text{and} \qquad -\sqrt{9} = -3$$

It is important to mention that taking the square root of a number and squaring a number are inverse operations. For example, 3 squared is written as 3^2, which equals 9. So, the square root of 9 equals 3 since $3 \times 3 = 3^2 = 9$.

Cube roots

To find the **cube root** of a number, you want to find some number that when multiplied by itself twice gives you the original number. For example, to find the cube root of 8, you want to find the number that when multiplied by itself twice gives you 8. The cube root of 8, then, is 2, because $2 \times 2 \times 2 = 8$. Following is a list of the first eleven cube roots of perfect cubes.

$$\sqrt[3]{0} = 0 \qquad \sqrt[3]{64} = 4 \qquad \sqrt[3]{512} = 8$$

$$\sqrt[3]{1} = 1 \qquad \sqrt[3]{125} = 5 \qquad \sqrt[3]{729} = 9$$

$$\sqrt[3]{8} = 2 \qquad \sqrt[3]{216} = 6 \qquad \sqrt[3]{1{,}000} = 10$$

$$\sqrt[3]{27} = 3 \qquad \sqrt[3]{343} = 7$$

As mentioned in the "Square roots" section, taking the square root of a number and squaring a number are inverse operations. Similarly, taking the cube root of a number and cubing a number are inverse operations. So, $2^3 = 8$ and $\sqrt[3]{8} = 2$. This understanding is used to solve equations that contain square or cube numbers. In Common Core Mathematics, square root and cube root symbols represent solutions to equations in the form $x^2 = y$ and $x^3 = y$ (y is a positive rational number). For example, if $x^2 = 9$, the solutions are square roots of 9 and are written as $x = \pm\sqrt{9}$; therefore, $x = \pm 3$. Note that there are two solutions to this equation since $(-3)(-3) = 9$ and $3(3) = 9$.

Approximating square roots

To find the square root of a number that is not a perfect square, it will be necessary to find an approximate answer. ***Note:*** Nonperfect square roots and nonperfect cube roots are irrational numbers.

Example 7: Find an approximation of $\sqrt{42}$.

Since $6^2 = 36$ and $7^2 = 49$, then $\sqrt{42}$ is between $\sqrt{36}$ and $\sqrt{49}$. Therefore, $\sqrt{42}$ is a value between 6 and 7. Since 42 is about halfway between 36 and 49, you can expect that $\sqrt{42}$ will be close to halfway between 6 and 7, or about 6.5. To check this estimation, $6.5 \times 6.5 = 42.25$, or about 42.

Another method that can be used to find a closer approximation is to first find the perfect square root closest to $\sqrt{42}$, which is $\sqrt{36}$. (Since $\sqrt{42}$ is closer to $\sqrt{36}$ than it is to $\sqrt{49}$, the approximation will be closer to 6 than to 7.) Then, divide 6 (the distance between 36 and 42) by 13 (the distance between the perfect square roots of 36 and 49) to get 0.46. Another estimate of $\sqrt{42}$ would be 6.46. (The actual square root is 6.48 rounded to the nearest hundredth.)

Square roots of nonperfect squares can be approximated, looked up in tables, or found using a calculator. You may want to keep these two in mind:

$$\sqrt{2} \approx 1.414 \qquad \sqrt{3} \approx 1.732$$

Simplifying square roots

Sometimes you will have to simplify nonperfect square roots, or write them in simplest form. In fractions, $\frac{2}{4}$ can be reduced to $\frac{1}{2}$. In square roots, $\sqrt{32}$ can be simplified to $4\sqrt{2}$.

There are two main methods to simplify a square root.

Method 1. Factor the radicand into a product of two factors, one of which being the largest possible perfect square. (Perfect squares are 1, 4, 9, 16, 25, 36, 49, …)

Method 2. Completely factor the radicand into a product of prime numbers (prime factorization) and then simplify by bringing out any factors that come in pairs.

Example 8: Simplify $\sqrt{32}$.

Method 1

$$\begin{aligned}\sqrt{32} &= \sqrt{16\times 2}\\ &= \sqrt{16}\times\sqrt{2}\\ &= 4\sqrt{2}\end{aligned}$$

Method 2

$$\begin{aligned}\sqrt{32} &= \sqrt{2\times 16}\\ &= \sqrt{2\times 2\times 8}\\ &= \sqrt{2\times 2\times 2\times 4}\\ &= \sqrt{2\times 2\times 2\times 2\times 2}\\ &= \sqrt{2\times 2}\times\sqrt{2\times 2}\times\sqrt{2}\\ &= 2\times 2\times\sqrt{2}\\ &= 4\sqrt{2}\end{aligned}$$

In Example 8, the largest perfect square is easy to see, so Method 1 is probably the faster method.

Example 9: Simplify $\sqrt{2{,}016}$.

Method 1

$$\begin{aligned}\sqrt{2{,}016} &= \sqrt{144\times 14}\\ &= \sqrt{144}\times\sqrt{14}\\ &= 12\sqrt{14}\end{aligned}$$

Method 2

$$\begin{aligned}\sqrt{2{,}016} &= \sqrt{2\times 1{,}008}\\ &= \sqrt{2\times 2\times 504}\\ &= \sqrt{2\times 2\times 2\times 252}\\ &= \sqrt{2\times 2\times 2\times 2\times 126}\\ &= \sqrt{2\times 2\times 2\times 2\times 2\times 63}\\ &= \sqrt{2\times 2\times 2\times 2\times 2\times 3\times 21}\\ &= \sqrt{2\times 2\times 2\times 2\times 2\times 3\times 3\times 7}\\ &= \sqrt{2\times 2}\times\sqrt{2\times 2}\times\sqrt{3\times 3}\times\sqrt{2\times 7}\\ &= 2\times 2\times 3\times\sqrt{14}\\ &= 12\sqrt{4}\end{aligned}$$

In Example 9, it is not so obvious that the largest perfect square is 144, so Method 2 is probably the faster method.

Many square roots cannot be simplified because they are already in simplest form, such as $\sqrt{7}$, $\sqrt{10}$, and $\sqrt{15}$.

Evaluating Numerical Expressions with Grouping Symbols

Common Core Standard: Seeing the Structure in Expressions

Interpret expressions that represent a quantity in terms of their context: (a) interpret parts of an expression, and (b) interpret complicated expressions by viewing one or more of their parts as a single entity (A.SSE.1).

Parentheses ( ), brackets [ ], and braces { } are frequently used to group numerical expressions. Operations inside grouping symbols must be performed before any operations outside of the grouping symbols.

Parentheses ( )

Parentheses are used to group numbers or variables. Everything inside parentheses must be done before any other operations, unless the distributive property is used.

Example 10: Simplify $50(2 + 6)$.

$$50(2+6) = 50(8) = 400$$

When a parenthesis is preceded by a minus sign, to remove the parentheses, change the sign of each term within the parentheses.

Example 11: Simplify $6 - (-3 + a - 2b + c)$.

$$\begin{aligned} 6-(-3+a-2b+c) &= \\ 6+3-a+2b-c &= \\ 9-a+2b-c \end{aligned}$$

Brackets [ ] and braces { }

Brackets and **braces** also are used to group numbers or variables. Technically, they are used after parentheses. Parentheses are to be used first, then brackets, and then braces: $\{[(\ )]\}$. Sometimes, instead of brackets or braces, you will see the use of larger parentheses.

$$((3+4)\cdot 5)+2$$

An expression using all three grouping symbols would look like this:

$$2\{1 + [4(2 + 1) + 3]\}$$

Example 12: Simplify $2\{1 + [4(2 + 1) + 3]\}$.

Notice that you work from the inside out.

$$\begin{aligned} 2\{1+[4(2+1)+3]\} &= \\ 2\{1+[4(3)+3]\} &= \\ 2\{1+[12+3]\} &= \\ 2\{1+[15]\} &= \\ 2\{16\} &= 32 \end{aligned}$$

Order of operations

When simplifying numerical expressions that contain two or more operations, there is a specific order in which these operations are to be performed. If multiplication, division, powers, addition, parentheses, and so forth are all contained in one problem, the order of operations is as follows:

1. Parentheses (or other grouping symbols)
2. Exponents (or radicals)
3. Multiplication or division (in the order it occurs from left to right)
4. Addition or subtraction (in the order it occurs from left to right)

Hint: The mnemonic PEMDAS will help you remember the order of operations: **P**lease **E**xcuse **M**y **D**ear **A**unt **S**ally (**P**arentheses, **E**xponents, **M**ultiplication, **D**ivision, **A**ddition, and **S**ubtraction). The "P" reminds you that "parentheses" are done first, the "E" reminds you that "exponents" are done next, the "MD" reminds you to "multiply or divide" in the order it occurs from left to right, and the "AS" reminds you to "add or subtract" in the order it occurs from left to right.

Example 13: Simplify the following numerical expressions.

(a) $6+4\times 3=$

$6+12=$ (multiplication)

18 (then addition)

(b) $10-3\times6+10^2+(6+1)\times4=$

$$10-3\times6+10^2+(7)\times4= \text{(parentheses first)}$$

$$10-3\times6+100+7\times4= \text{(exponents next)}$$

$$10-18+100+28= \text{(multiplication)}$$

$$-8+100+28= \text{(addition/subtraction left to right)}$$

$$92+28=120$$

Divisibility Rules

Common Core Mathematics expects that you can fluently compute multi-digit numbers and find common factors and multiples. The following set of rules can help you save time in trying to check the divisibility of multi-digit numbers.

A Number is Divisible by	*If*
2	it ends in 0, 2, 4, 6, or 8.
3	the sum of its digits is divisible by 3.
4	the number formed by the last two digits is divisible by 4.
5	it ends in 0 or 5.
6	it is divisible by 2 and 3 (use the rules for both).
7	(no simple rule).
8	the number formed by the last three digits is divisible by 8.
9	the sum of its digits is divisible by 9.
10	it ends in 0.

Example 14:

(a) Is 126 divisible by 3? Sum of digits = 9. Because 9 is divisible by 3, then 126 is divisible by 3.

(b) Is 1,648 divisible by 4? Because 48 is divisible by 4, then 1,648 is divisible by 4.

(c) Is 186 divisible by 6? Because 186 ends in 6, it is divisible by 2. Sum of digits = 15. Because 15 is divisible by 3, 186 is divisible by 3. 186 is divisible by 2 and 3; therefore, it is divisible by 6.

(d) Is 2,488 divisible by 8? Because 488 is divisible by 8, then 2,488 is divisible by 8.

(e) Is 2,853 divisible by 9? Sum of digits = 18. Because 18 is divisible by 9, then 2,853 is divisible by 9.

Signed Numbers (Positive Numbers and Negative Numbers)

The term *signed numbers* refers to positive and negative numbers. If no sign is shown, the number automatically is considered positive.

Number lines

Common Core Mathematics stresses the importance of understanding that signed numbers can be visually represented on a horizontal number line. On a **number line,** numbers to the right of 0 are positive. Numbers to the left of 0 are negative, as shown in the figure below.

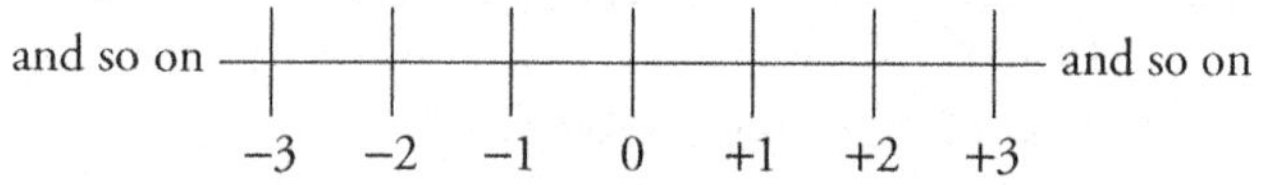

Given any two numbers on a number line, the one on the right is always larger, regardless of its sign (positive or negative). Note that fractions may also be placed on a number line, as shown in the figure below. When comparing real numbers, using a number line gives a visual representation of which one is of lesser or of greater value. For example, $-8 < -4$ since -8 is located to the left of -4 on a number line.

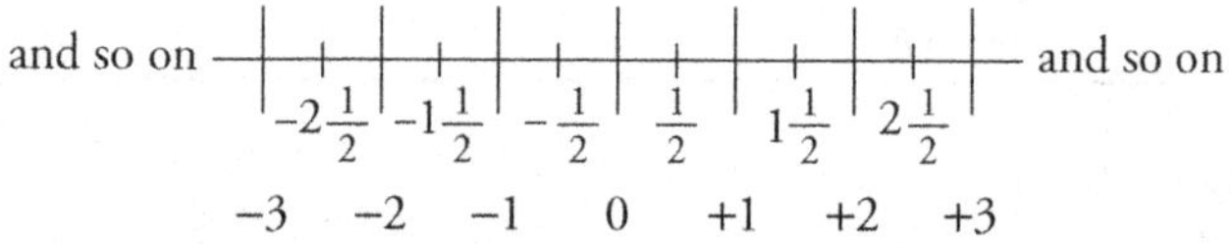

Absolute Value

Common Core Mathematics connects the concept of absolute value to the addition and subtraction of signed numbers on a number line. (The topic of absolute value is discussed in Chapter 7, "Inequalities, Graphing, and Absolute Value.") A number's distance from zero on a number line is called **absolute value.** For any number, *n,* the absolute value of *n* is denoted as $|n|$. Since distance is always nonnegative, the absolute value is always nonnegative. For example, $|5| = 5$ and $|-5| = 5$ since 5 and -5 are both five units away from 0 on the number line.

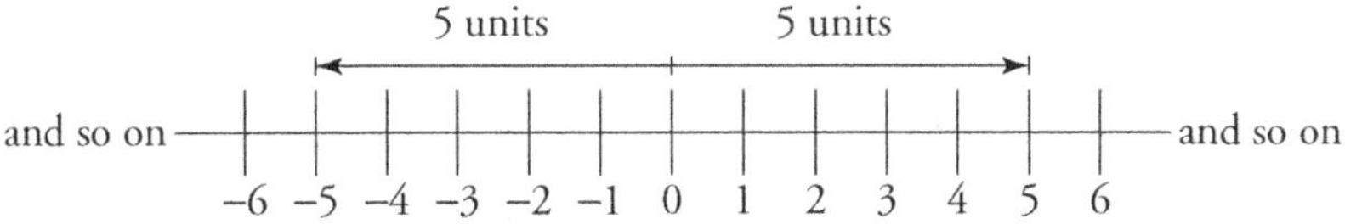

Note: You can use absolute value as a rule to add and subtract signed numbers.

Addition of signed numbers

When adding two numbers with the same sign (either both positive or both negative), add their absolute values. The sum will have the same sign as the addends. Addition problems can be presented in either a vertical form (up and down) or a horizontal form (across).

Example 15: Find each sum.

(a) $\begin{array}{r} +5 \\ \underline{+(+7)} \\ ? \end{array}$ Add the absolute values $(5+7=12)$ and keep the sign $(+)$.

$$\begin{array}{r} +5 \\ \underline{+(+7)} \\ +12 \end{array}$$

(b) $-8+(-3)=?$ Add the absolute values $(8+3=11)$ and keep the sign $(-)$.

$-8+(-3)=-11$

When adding two numbers with different signs (one positive and one negative), subtract their absolute values. The sum will have the same sign of the addend with the larger absolute value.

Example 16: Find each sum.

(a)
$$\begin{array}{r} +5 \\ +(-7) \\ \hline ? \end{array}$$

Subtract the absolute values $(7-5=2)$ and keep the sign of the addend with the larger absolute value (-7).

$$\begin{array}{r} +5 \\ +(-7) \\ \hline -2 \end{array}$$

(b) $-59+(+72)=?$

Subtract the absolute values $(72-59=13)$ and keep the sign of the addend with the larger absolute value $(+72)$.

$-59+(+72)=+13$

Example 17: Find each sum.

(a) $\begin{array}{r} +9 \\ +(+6) \\ \hline \end{array}$ **(b)** $-12+9$ **(c)** $\begin{array}{r} 8 \\ +(-5) \\ \hline \end{array}$ **(d)** $23+(-18)$

$\begin{array}{r} +9 \\ +(+6) \\ \hline +15 \end{array}$ $-12+9=-3$ $\begin{array}{r} 8 \\ +(-5) \\ \hline +3 \end{array}$ $23+(-18)=5$

Subtraction of signed numbers

To subtract positive and/or negative numbers, add the additive inverse (the opposite) of the number being subtracted. For every real number, a and b, $a-b=a+(-b)$.

Example 18: Find each difference.

(a) $\begin{array}{r} +12 \\ -(+4) \\ \hline \end{array}$ **(b)** $\begin{array}{r} -14 \\ -(-4) \\ \hline \end{array}$ **(c)** $\begin{array}{r} -19 \\ -(+6) \\ \hline \end{array}$ **(d)** $\begin{array}{r} +20 \\ -(-3) \\ \hline \end{array}$

$\begin{array}{r} +12 \\ +(-4) \\ \hline +8 \end{array}$ $\begin{array}{r} -14 \\ +(+4) \\ \hline -10 \end{array}$ $\begin{array}{r} -19 \\ +(-6) \\ \hline -25 \end{array}$ $\begin{array}{r} +20 \\ +(+3) \\ \hline +23 \end{array}$

Subtracting positive and/or negative numbers may also be done "horizontally."

Example 19: Find each difference.

(a) $+12 - (+4) = +12 + (-4) = 8$

(b) $+16 - (-6) = +16 + (+6) = 22$

(c) $-20 - (+3) = -20 + (-3) = -23$

(d) $-5 - (-2) = -5 + (+2) = -3$

Minus sign preceding parenthesis

If a minus sign precedes a parenthesis, it means everything within the parentheses is to be subtracted. Therefore, using the same rule as in subtraction of signed numbers, simply change every sign within the parentheses to its additive inverse and then add.

Example 20: Simplify the following.

(a) $9-(+3-5+7-6)=$

$9+(-3+5-7+6)=$

$9+(+1)=10$

(b) $20-(+35-50+100)=$

$20+(-35+50-100)=$

$20+(-85)=-65$

Multiplying and dividing signed numbers

To multiply or divide signed numbers, treat them just like regular numbers, but remember this rule: An odd number of negative signs will produce a negative answer. An even number of negative signs will produce a positive answer.

Example 21: Find each product or quotient.

(a) $(-3)(+8)(-5)(-1)(-2)=+240$

(b) $(-3)(+8)(-1)(-2)=-48$

(c) $\dfrac{-64}{-2}=+32$

(d) $\dfrac{-64}{+2}=-32$

Fractions

A **fraction,** or fractional number, is used to represent a part of a whole. Common Core Mathematics states that you must interpret and compute quotients of fractions and solve fraction-related word problems using visual fraction models and equations (understand that a fraction $\frac{a}{b}$ is the quantity formed by *a* parts of size $\frac{1}{b}$).

Fractions consist of two numbers: a **numerator** (which is above the fraction bar) and a **denominator** (which is below the fraction bar).

$$\frac{1}{2}\ \frac{\text{numerator}}{\text{denominator}}$$

The denominator represents the number of equal parts that the whole is being divided into. The numerator represents the number of equal parts that are being considered. Thus, if the fraction is $\frac{3}{5}$, or three-fifths of a pie, the denominator 5 tells you that the pie has been divided into five equal parts, of which 3 (numerator) are being considered. Sometimes, it helps to think of the dividing line (the middle of a fraction) as meaning "out of." In other words, $\frac{3}{5}$ also means 3 out of 5 equal parts from the whole pie.

Negative fractions

Fractions may be negative as well as positive. (See the number line on p. 24.) However, negative fractions are typically written as follows:

$$-\frac{3}{4} \text{ not } \frac{-3}{4} \text{ or } \frac{3}{-4} \text{ (although they are all equal)}$$

$$-\frac{3}{4} = \frac{-3}{4} = \frac{3}{-4}$$

Adding and subtracting fractions

When you add or subtract fractions with like denominators, find the sum or difference of the numerators and leave the denominators unchanged. When you add or subtract fractions with unlike denominators, write them in an equivalent form (as equivalent fractions) with like denominators before adding or subtracting. Write the answer in simplest form (reduce the fraction to lowest terms) if necessary.

Note: The rules for adding signed numbers apply to fractions as well.

Example 22: Find the sum or difference of the following.

(a) $-\frac{1}{2}+\frac{1}{3}=-\frac{3}{6}+\frac{2}{6}=-\frac{1}{6}$

(b)

$$\begin{array}{rcr} +\frac{3}{4} & = & +\frac{9}{12} \\ \underline{+\left(-\frac{1}{3}\right)} & = & \underline{+\left(-\frac{4}{12}\right)} \\ & & +\frac{5}{12} \end{array}$$

(c)

$$\begin{array}{rcrcr} +\frac{9}{10} & = & +\frac{9}{10} & = & +\frac{9}{10} \\ \underline{-\left(-\frac{1}{5}\right)} & = & \underline{+\frac{1}{5}} & = & \underline{+\frac{2}{10}} \\ & & & & +\frac{11}{10}=1\frac{1}{10} \end{array}$$

(d) $+\frac{2}{3}-\left(-\frac{1}{5}\right)=\frac{10}{15}-\left(-\frac{3}{15}\right)=\frac{10}{15}+\left(+\frac{3}{15}\right)=\frac{13}{15}$

(e) $+\frac{1}{3}-\frac{3}{4}=+\frac{4}{12}-\frac{9}{12}=+\frac{4}{12}+\left(-\frac{9}{12}\right)=-\frac{5}{12}$

Multiplying fractions

To multiply fractions, simply multiply the numerators and then multiply the denominators. Reduce the answer to simplest form (simplify) if necessary.

Example 23: Find the product.

$$\frac{2}{3}\times\frac{5}{12}=\frac{10}{36} \qquad \text{simplify } \frac{10}{36} \text{ to } \frac{5}{18}$$

This answer has to be reduced because it isn't in simplest form. Because whole numbers can also be written as fractions $\left(3=\frac{3}{1},\ 4=\frac{4}{1}, \text{ and so forth}\right)$, the problem $3\times\frac{3}{8}$ would be worked by changing 3 to $\frac{3}{1}$. So, $\frac{3}{1}\times\frac{3}{8}=\frac{9}{8}$.

Simplifying (reducing) fractions

Simplifying (reducing) common factors is a method that can be used to make multiplying fractions easier to work with. Instead of simplifying your answer after multiplying, you can *cross-cancel* first, by simplifying fractions before multiplying. To simplify, find a common factor that divides evenly into one numerator and one denominator. In this case, 2 divides into both 2, the numerator, and 12, the denominator, evenly. Thus, you can cross-cancel as follows:

$$\frac{\overset{1}{\cancel{2}}}{3} \times \frac{5}{\underset{6}{\cancel{12}}} = \frac{5}{18}$$

Remember, you may simplify fractions only when multiplying them, and it can be done either diagonally top to bottom or vertically top to bottom. The rules for multiplying signed numbers hold here, too.

Example 24: Simplify first, then find the product.

(a) $\frac{1}{4} \times \frac{2}{7} = \frac{1}{\underset{2}{\cancel{4}}} \times \frac{\overset{1}{\cancel{2}}}{7} = \frac{1}{14}$

(b) $\left(-\frac{3}{8}\right) \times \left(-\frac{4}{9}\right) = \left(-\frac{\overset{1}{\cancel{3}}}{\underset{2}{\cancel{8}}}\right) \times \left(-\frac{\overset{1}{\cancel{4}}}{\underset{3}{\cancel{9}}}\right) = +\frac{1}{6}$

Multiplying mixed numbers

To multiply mixed numbers, first change any mixed number to an improper fraction. Then multiply as shown earlier in this chapter (see "Multiplying fractions").

Example 25: Find the product.

$$3\frac{1}{3} \times 2\frac{1}{4} = \frac{10}{3} \times \frac{9}{4} = \frac{\overset{5}{\cancel{10}}}{\underset{1}{\cancel{3}}} \times \frac{\overset{3}{\cancel{9}}}{\underset{2}{\cancel{4}}} = \frac{15}{2} \text{ or } 7\frac{1}{2}$$

Change the answer, if in improper fraction form, back to a mixed number and simplify if necessary. Remember, the rules for multiplication of signed numbers apply here as well.

Dividing fractions

To divide fractions, change the division problem to a multiplication problem by taking the reciprocal or multiplicative inverse (turn upside down) of the second fraction (the one "divided by"). Then multiply the fractions and simplify if possible.

Example 26: Find the quotient.

(a) $\frac{1}{6} \div \frac{1}{5} = \frac{1}{6} \times \frac{5}{1} = \frac{5}{6}$

(b) $\frac{1}{9} \div \frac{1}{3} = \frac{1}{\cancel{9}_3} \times \frac{\cancel{3}^1}{1} = \frac{1}{3}$

Here, too, the rules for division of signed numbers apply.

Dividing complex fractions

Sometimes a division of fractions problem may appear as a **complex fraction,** where a fraction contains one or more fractions in its numerator, in its denominator, or in both.

Example 27: Simplify $\dfrac{\frac{3}{4}}{\frac{7}{8}}$.

Consider the line separating the two fractions to mean "divided by." Therefore, this problem can be rewritten in equivalent form as follows:

$$\frac{3}{4} \div \frac{7}{8}$$

Now, follow the same procedure as shown in Example 26.

$$\frac{3}{4} \div \frac{7}{8} = \frac{3}{\cancel{4}_1} \times \frac{\cancel{8}^2}{7} = \frac{6}{7}$$

Dividing mixed numbers

To divide mixed numbers, first change them to improper fractions (see the "Multiplying mixed numbers" section earlier in this chapter). Then follow the rule for dividing fractions (see the "Dividing fractions" section earlier in this chapter).

Example 28: Find the quotient.

$$3\frac{3}{5} \div 2\frac{2}{3} = \frac{18}{5} \div \frac{8}{3} = \frac{\overset{9}{\cancel{18}}}{5} \times \frac{3}{\underset{4}{\cancel{8}}} = \frac{27}{20} \text{ or } 1\frac{7}{20}$$

Notice that after you invert and have a multiplication of fractions problem, you then may cancel when appropriate.

Simplifying fractions and complex fractions

Common Core Mathematics suggests that you extend your understanding of the properties of operations as you simplify fractions. If either the numerator or the denominator contains a numerical expression, simplify first before you find the quotient. Then reduce if possible.

Example 29: Simplify each complex fraction.

(a) $\frac{28+14}{26+17} = \frac{42}{43}$

(b) $\dfrac{\frac{1}{4}+\frac{1}{2}}{\frac{1}{3}+\frac{1}{4}} = \dfrac{\frac{1}{4}+\frac{2}{4}}{\frac{4}{12}+\frac{3}{12}} = \dfrac{\frac{3}{4}}{\frac{7}{12}} = \frac{3}{4} \div \frac{7}{12} = \frac{3}{\underset{1}{\cancel{4}}} \times \frac{\overset{3}{\cancel{12}}}{7} = \frac{9}{7} \text{ or } 1\frac{2}{7}$

(c) $\dfrac{3-\frac{3}{4}}{-4+\frac{1}{2}} = \dfrac{2\frac{1}{4}}{-3\frac{1}{2}} = \dfrac{\frac{9}{4}}{-\frac{7}{2}} = \frac{9}{4} \div \left(-\frac{7}{2}\right) = \frac{9}{\underset{2}{\cancel{4}}} \times \left(-\frac{\overset{1}{\cancel{2}}}{7}\right) = -\frac{9}{14}$

(d) $$\frac{1}{1+\frac{1}{1+\frac{1}{4}}}=\frac{1}{1+\frac{1}{\frac{5}{4}}}=\frac{1}{1+\left(1\div\frac{5}{4}\right)}=\frac{1}{1+\left(1\times\frac{4}{5}\right)}=\frac{1}{1+\frac{4}{5}}=$$

$$\frac{1}{1\frac{4}{5}}=\frac{1}{\frac{9}{5}}=1\div\frac{9}{5}=1\times\frac{5}{9}=\frac{5}{9}$$

Decimals

Common Core Mathematics confirms that every rational number can be expressed as a fraction. Fractions can be converted to decimals, as either terminating, coming to an end (for example, 0.3 or 0.125), or as having an **infinite** (never-ending) repeating pattern (for example, (0.666 … or 0.1272727…).

Converting terminating decimals to fractions

To convert terminating decimals to fractions, use the concept of place value, in that all numbers to the right of the decimal point are fractions with denominators written as a power of 10. Next, use the technique of *read it, write it,* and *reduce it.*

Example 30: Change each decimal to a fraction. Simplify when possible.

(a) 0.8

Read it: eight-tenths

Write it: $\frac{8}{10}$

Reduce it: $\frac{4}{5}$

(b) -0.07

Read it: negative seven-hundredths

Write it: $-\frac{7}{100}$ (can't reduce this one)

All rules for signed numbers also apply to operations with decimals.

Converting fractions to decimals

To convert a fraction to a decimal, use long division to divide the numerator by the denominator. For example, $\frac{13}{20}$ means 13 divided by 20. The quotient will be a terminating decimal or a repeating decimal.

Example 31: Change each fraction to a decimal.

(a) $\frac{13}{20}$

$$20\overline{)13.00}^{\;0.65} = 0.65$$

(b) $-\frac{2}{9}$

$$9\overline{)-2.00000}^{\;-0.222\ldots} = \;0.222\ldots$$

Converting repeating decimals to fractions

Repeating decimals usually are represented by putting a line over (sometimes under) the shortest block of repeating decimals. This line is called a **vinculum.** So you would write:

$0.\overline{3}$ to indicate .333...

$0.\overline{51}$ to indicate .515151...

$-2.1\overline{47}$ to indicate $-2.1474747\ldots$

Notice that only the digits under the vinculum are repeated.

Every repeating decimal can be expressed as a fraction using either patterns or algebraic reasoning. The examples below use algebraic reasoning to find the fraction equivalent of a repeating decimal. Example 32 thoroughly explains the process.

Example 32: Change $0.\overline{7}$ to a fraction.

First, write an equation that represents the repeating decimal that you want to convert to a fraction. Then, multiply both sides of the equation by a power of 10 so that both equations have the same repeating digits to the right of the decimal point.

Let $n = 0.\overline{7}$ or $0.7777777\ldots$

Let $10n = 7.\overline{7}$ or $7.777777\ldots$

Because $10n$ and n have the same repeating digit to the right of the decimal point, their difference is an integer. Subtract the original equation from the new equation and solve the equation to determine the equivalent fraction.

$$\begin{aligned} 10n &= 7.\overline{7} \\ \underline{-n} &= \underline{-0.\overline{7}} \\ 9n &= 7 \\ n &= \frac{7}{9} \end{aligned}$$

$$\text{Therefore, } 0.\overline{7} = \frac{7}{9}.$$

When fractions have denominators of 9, 99, and 11, repeating patterns occur. Let's explore how to determine the fraction equivalent of the example above using the patterns below.

$$\frac{1}{9} = 0.\overline{1} \qquad \frac{4}{9} = 0.\overline{4}$$

$$\frac{2}{9} = 0.\overline{2} \qquad \frac{5}{9} = 0.\overline{5}$$

$$\frac{3}{9} = 0.\overline{3} \qquad \text{and so on...}$$

Based on the pattern above, we can conclude that $0.\overline{7} = \frac{7}{9}$.

Example 33: Change $0.\overline{36}$ to a fraction.

Let $n = 0.\overline{36}$ or $0.363636\ldots$

Let $100n = 36.\overline{36}$ or $36.363636\ldots$

Because $100n$ and n have the same repeating digits to the right of the decimal point, their difference is an integer.

$$\begin{aligned} 100n &= 36.\overline{36} \\ -n &= -0.\overline{36} \\ 99n &= 36 \\ n &= \frac{36}{99} = \frac{4}{11} \end{aligned}$$

$$\text{Therefore, } 0.\overline{36} = \frac{4}{11}.$$

Example 34: Change $0.5\overline{4}$ to a fraction.

Let $n = 0.5\overline{4}$ or 0.54444…

Let $10n = 5.\overline{4}$ or 5.4444…

Let $100n = 54.\overline{4}$ or 54.4444…

Because $100n$ and $10n$ have the same repeating digit to the right of the decimal point, their difference is an integer.

$$\begin{aligned} 100n &= 54.\overline{4} \\ -10n &= -5.\overline{4} \\ 90n &= 49 \\ n &= \frac{49}{90} \end{aligned}$$

$$\text{Therefore, } 0.5\overline{4} = \frac{49}{90}.$$

Percents, Fractions, and Proportions

Common Core Mathematics requires that you recognize and represent proportional relationships between quantities. To help you identify proportional relationships, this section will cover arithmetic percents, fractions, and proportions.

Percents

A fraction whose denominator is 100 is called a **percent.** The word *percent* means parts (or rate) per one-hundred, which is related to part-to-whole ratios, and can be extended to proportional reasoning. Common

Core Mathematics frequently uses percents to solve real-world mathematical problems involving ratio and proportional relationships. For example, if you take 37% of a bag of seeds, that would be equivalent to dividing the seeds into 100 equal parts, and then taking 37 of those parts. So, 37% is equivalent to thirty-seven hundredths, which can be written as a fraction and as a decimal.

$$37\% = \frac{37}{100} = 0.37$$

Converting decimals to percents

To convert decimals to percents, take the following steps:

1. Multiply the decimal by 100, or move the decimal point two places to the right.
2. Insert a percent sign.

Example 35: Convert each decimal to a percent.

(a) 0.09 = 9%

(b) 0.75 = 75%

(c) 1.85 = 185%

(d) 0.002 = 0.2%

(e) 8.7 = 870%

Converting percents to decimals

To convert percents to decimals, take the following steps:

1. Eliminate the percent sign.
2. Divide the number by 100, or move the decimal point two places to the left (sometimes, adding zeros will be necessary).

Example 36: Convert each percent to a decimal.

(a) 23% = 0.23

(b) 5% = 0.05

(c) 0.7% = 0.007

(d) $16\frac{2}{3}\% = 0.16\frac{2}{3}$

Converting fractions to percents

There are two simple methods for converting fractions to percents.

Method 1.

1. Change the fraction to a decimal.
2. Change the decimal to a percent.

Example 37 (applying Method 1): Convert each fraction to a percent.

(a) $\frac{2}{5} = 0.4 = 40\%$

(b) $\frac{5}{2} = 2.5 = 250\%$

(c) $\frac{1}{20} = 0.05 = 5\%$

Method 2.

1. Write a proportion (a statement where two ratios are equal) with the fraction on the left side of the equal sign and $\frac{x}{100}$ on the right side of the equal sign, where x is the percent.
2. Solve the proportion for x, then place a % sign to its right.

Example 38 (applying Method 2): Convert each fraction to a percent.

(a) $\frac{3}{4} \rightarrow \frac{3}{4} = \frac{x}{100}$

$$4x = 300$$

$$x = \frac{300}{4} = 75 \qquad \text{Therefore, } \frac{3}{4} = 75\%.$$

(b) $\frac{2}{3} \rightarrow \frac{2}{3} = \frac{x}{100}$

$$3x = 200$$

$$x = \frac{200}{3} = 66\frac{2}{3} \qquad \text{Therefore, } \frac{2}{3} = 66\frac{2}{3}\%.$$

(c) $\frac{1}{20} \rightarrow \frac{1}{20} = \frac{x}{100}$

$$20x = 100$$

$$x = \frac{100}{20} = 5 \qquad \text{Therefore, } \frac{1}{20} = 5\%.$$

Converting percents to fractions

There are two simple methods for converting percents to fractions.

Method 1.

1. Drop the percent sign.
2. Write over 100.
3. Simplify if necessary.

Example 39 (applying Method 1): Convert each percent to a fraction.

(a) $60\% = \frac{60}{100} = \frac{3}{5}$

(b) $230\% = \frac{230}{100} = \frac{23}{10} \left(\text{or } 2\frac{3}{10}\right)$

Method 2.

1. Drop the percent sign.
2. Multiply by $\frac{1}{100}$.
3. Simplify if necessary.

Example 40 (applying Method 2): Convert each percent to a fraction.

(a) $66\frac{2}{3}\% \rightarrow 66\frac{2}{3} \times \frac{1}{100} = \frac{\overset{2}{\cancel{200}}}{3} \times \frac{1}{\underset{1}{\cancel{100}}} = \frac{2}{3}$

(b) $112\frac{1}{2}\% \rightarrow 112\frac{1}{2} \times \frac{1}{100} = \frac{\overset{9}{\cancel{225}}}{2} \times \frac{1}{\underset{4}{\cancel{100}}} = \frac{9}{8} \text{ or } 1\frac{1}{8}$

Memorizing the following conversion can eliminate computations:

$\frac{1}{100} = 0.01 = 1\%$

$\frac{9}{10} = 0.9 = 90\%$

$\frac{2}{6} = \frac{1}{3} = 0.33\frac{1}{3} = 33\frac{1}{3}\%$

$\frac{1}{10} = 0.1 = 10\%$

$\frac{1}{8} = 0.125 = 12.5\% = 12\frac{1}{2}\%$

$\frac{3}{6} = \frac{1}{2} = 0.5 = 50\%$

$\frac{2}{10} = \frac{1}{5} = 0.2 = 20\%$

$\frac{2}{8} = \frac{1}{4} = 0.25 = 25\%$

$\frac{4}{6} = \frac{2}{3} = 0.66\frac{2}{3} = 66\frac{2}{3}\%$

$\frac{3}{10} = 0.3 = 30\%$

$\frac{3}{8} = 0.375 = 37.5\% = 37\frac{1}{2}\%$

$\frac{5}{6} = 0.83\frac{1}{3} = 83\frac{1}{3}\%$

$\frac{4}{10} = \frac{2}{5} = 0.4 = 40\%$

$\frac{4}{8} = \frac{1}{2} = 0.5 = 50\%$

$1 = 1.0 = 100\%$

$\frac{5}{10} = \frac{1}{2} = 0.5 = 50\%$

$\frac{5}{8} = 0.625 = 62.5\% = 62\frac{1}{2}\%$

$2 = 2.0 = 200\%$

$\frac{6}{10} = \frac{3}{5} = 0.6 = 60\%$

$\frac{6}{8} = \frac{3}{4} = 0.75 = 75\%$

$3\frac{1}{2} = 3.5 = 350\%$

$\frac{7}{10} = 0.7 = 70\%$

$\frac{7}{8} = 0.875 = 87.5\% = 87\frac{1}{2}\%$

$\frac{8}{10} = \frac{4}{5} = 0.8 = 80\%$

$\frac{1}{6} = 0.16\frac{2}{3} = 16\frac{2}{3}\%$

Finding the percent of a quantity

To determine the percent of a quantity, change the percent to a fraction or decimal (whichever is easier for you) and multiply. Remember, the word *of* means multiply.

Example 41: Find the percent of each quantity.

(a) 20% of 80 =

$$\frac{\overset{1}{\cancel{20}}}{\underset{5}{\cancel{100}}} \times \frac{80}{1} = \frac{80}{5} = 16 \text{ or } 0.20 \times 80 = 16.00 = 16$$

(b) $\frac{1}{2}$% of 18 =

$$\frac{\frac{1}{2}}{100} \times \frac{18}{1} = \frac{1}{\underset{100}{\cancel{200}}} \times \frac{\overset{9}{\cancel{18}}}{1} = \frac{9}{100} \text{ or } 0.005 \times 18 = 0.09$$

Other applications of percent

You can solve problems involving percent, part, and whole using either an **equation** or a **proportion.** We will explore both methods below, and will use these methods to solve real-world problems involving percent in Chapter 12, "Application Problems."

The percent equation

Turn the question word-for-word into an equation. For *what*, substitute the letter *x*; for *is*, substitute an *equal sign;* for *of*, substitute a *multiplication sign.* Change percents to decimals or fractions, whichever you find easier. Then solve the equation.

Example 42: Determine whether you are finding the percent, part, or whole. Then, write an equation and solve.

(a) 18 is what percent of 90? (finding the percent)

$$18 = (x)(90)$$

$$\frac{18}{90} = x$$

$$\frac{1}{5} = x$$

$20\% = x$ Therefore, 18 is 20% of 90.

(b) 10 is 50 percent of what number? (finding the whole)

$$10 = (0.50)(x)$$
$$\frac{10}{0.50} = x$$
$$20 = x$$

Therefore, 10 is 50% of 20.

(c) What is 15 percent of 60? (finding the part)

$$x = \frac{15}{100} \times \frac{60}{1} = 9 \qquad \text{or} \qquad x = 0.15(60) = 9$$

Therefore, 9 is 15% of 60.

Using the method above leads us to the *percent equation,*

$$\text{percent} \times \text{whole} = \text{part}$$

where the percent is either written as a decimal or a rate per hundred.

The percent proportion

Although Common Core Mathematics uses a conceptual method to approach percent proportion problems (that identifies the part, whole, or percent before setting up the proportion), a quick method used to solve percent problems is the *is/of method.*

First set up a blank proportion and then fill in the empty spaces as illustrated below. Note that in this method, "is" is considered the "part" and "of" is considered the "whole."

$$\frac{\%\text{-number}}{100} = \frac{\text{"is"-number}}{\text{"of"-number}}$$

The is/of method works for the three basic types of percent questions:

- 30 is <u>what percent</u> of 50?
- 30 is 20% of <u>what number</u>?
- <u>What number</u> is 30% of 50?

Example 43: 30 is what percent of 50?

The "is" number is 30. The % number is unknown, so call it *x.*

The "of" number is 50. Now substitute these values in the appropriate positions in the proportion and solve for *x*.

$$\frac{x}{100} = \frac{30}{50}$$

In this particular instance, it can be observed that $\frac{60}{100} = \frac{30}{50}$, so the answer is 60%. Solving this problem mechanically would not be time-effective.

Example 44: 30 is 20% of what number?

$$\frac{20}{100} = \frac{30}{x}$$
$$20x = 3{,}000$$
$$x = 150$$

Therefore, 30 is 20% of 150.

Example 45: What number is 30% of 50?

$$\frac{30}{100} = \frac{x}{50}$$
$$100x = 1{,}500$$
$$x = 15$$

Therefore, 15 is 30% of 50.

Scientific Notation

Very large or very small quantities are sometimes written in **scientific notation.** Common Core Mathematics identifies operations with numbers expressed in scientific notation, including problems where both decimal and scientific notation are used.

For example, the mass of the moon is 73,476,730,900,000,000,000,000 kilograms. This weight can be hard to read and write correctly. Scientific notation can be used to simplify the quantity in the format $7.34767309 \times 10^{22}$ kilograms. A number in scientific notation is written as $a \times 10^n$, where $1 \leq a < 10$ and n is an integer.

Example 46: Express the following in scientific notation.

(a) 2,100,000

2,100,000 written in scientific notation is 2.1×10^6. Simply place the decimal point to get a decimal number greater than or equal to 1, but less than 10, and then count the number of digits from the original decimal point to the new one to get the power of 10.

2.100000. moved 6 digits to the left

(b) 0.0000004

0.0000004 written in scientific notation is 4.0×10^{-7}. Simply place the decimal point to get a decimal number greater than or equal to 1, but less than 10, and then count the number of digits from the original decimal point to the new one.

.0000004. moved 7 digits to the right

Notice that number values greater than 1 have positive exponents as the power of 10 and that number values between 0 and 1 have negative exponents as the power of 10.

Multiplication in scientific notation

To multiply numbers in scientific notation, multiply the decimal numbers together and then multiply the powers by keeping the base number 10 and adding the exponents. It may be necessary to make adjustments to this answer in order to correctly express it in scientific notation.

Example 47: Simplify each expression. Write each answer in scientific notation.

(a) $(2\times10^2)(3\times10^4)=(2\times3)\times10^{2+4}=6\times10^6$

(b) $(6\times10^5)(5\times10^7)=(6\times5)\times10^{5+7}=30\times10^{12}$

This answer must be changed to scientific notation (first number between 1 and 10):

$$30 \times 10^{12} = 3.0 \times 10^{1} \times 10^{12} = 3 \times 10^{13}$$

(c) $\left(4 \times 10^{-4}\right)\left(2 \times 10^{5}\right) = (4 \times 2) \times 10^{-4+5} = 8 \times 10^{1}$

Division in scientific notation

To divide numbers in scientific notation, divide the decimal numbers and then divide the powers by keeping the base number 10 and subtracting the exponents. It may be necessary to make adjustments to this answer in order to correctly express it in scientific notation.

Example 47: Simplify each expression. Write each answer in scientific notation.

(a) $\left(8 \times 10^{5}\right) \div \left(2 \times 10^{2}\right) = (8 \div 2) \times 10^{5-2} = 4 \times 10^{3}$

(b) $\dfrac{7 \times 10^{9}}{4 \times 10^{3}} = (7 \div 4) \times 10^{9-3} = 1.75 \times 10^{6}$

(c) $\left(6 \times 10^{7}\right) \div \left(3 \times 10^{9}\right) = (6 \div 3) \times 10^{7-9} = 2 \times 10^{-2}$

(d) $\left(2 \times 10^{4}\right) \div \left(5 \times 10^{2}\right) = (2 \div 5) \times 10^{4-2} = 0.4 \times 10^{2}$

This answer must be changed to scientific notation.

$$0.4 \times 10^{2} = 4 \times 10^{-1} \times 10^{2} = 4 \times 10^{1}$$

$$\text{(e)}\ \left(8.4\times10^{5}\right)\div\left(2.1\times10^{-4}\right)=(8.4\div2.1)\times10^{5-(-4)}=4\times10^{9}$$

Chapter Check-Out

Questions

1. Which of the following numbers are integers?

   $3,\ 4,\ \frac{1}{2},\ 0,\ -2, -5.8,\ \sqrt{6}$

2. How would you classify each of the following numbers that are not integers?

   $3,\ 4,\ \frac{1}{2},\ 0, -2, -5.8,\ \sqrt{6}$

3. Which of the following numbers are prime numbers?

   2, 5, 7, 9, 15, 21

4. The identity element in addition is ______.

5. Which of the following pairs of expressions is equivalent?

   A. $a - b$ and $b - a$
   B. $a(b + c)$ and $(a)(b)(c)$
   C. $4(a - 2b)$ and $4a - 2b$
   D. $3(a + c)$ and $a + c + a + c + a + c$

6. Simplify the expression $\left(\frac{1}{5}\right)^{-2}\cdot(-5)^{0}$.

7. Which of the following expressions is NOT equivalent to $\frac{3^4}{3^8}$?

A. $\frac{1}{3^2}$

B. $\frac{1}{3^4}$

C. 3^{-4}

D. $\frac{1}{81}$

8. Simplify the numerical expression $4^5 \times 4^8$ to an equivalent expression in exponential notation.

9. Simplify the numerical expression $\frac{2^{-1}}{16}$ to an equivalent expression in exponential notation.

10. Simplify the numerical expression $(3^2)^3$ to an equivalent expression in exponential notation.

11. Which of the following inequalities is NOT true?

A. $\sqrt{25} > \sqrt[3]{64}$

B. $\sqrt{42} \geq \sqrt[3]{216}$

C. $\sqrt[3]{1{,}000} \leq \sqrt{100}$

D. $\sqrt[3]{8} < \sqrt{4}$

12. Find an approximation of $\sqrt{54}$ to the nearest tenth.

Question 13 refers to the following information.

Lucas evaluated the numerical expression $3\left[10-\left(6+2^3 \div 4\times 2\right)\right]$ as follows:

$$3\left[10-\left(6+2^3 \div 4\times 2\right)\right]=$$
$$3[10-(6+8\div 4\times 2)]=$$
$$3[10-(6+8\div 8)]=$$
$$3[10-(6+1)]=$$
$$3[10-7]=$$
$$3(3)=9$$

13. Explain what Lucas did wrong and find the solution to correct his mistake.

14. The number 6,321 is evenly divisible by which numbers between 1 and 10?

15. The area of a square is 54 in^2. Find the length of the side of the square in simplest radical form.

For questions 16–19, select one of the following three answer choices to determine whether the statement is:

A. always true
B. sometimes true
C. never true

Explain your answers.

16. An integer is a rational number.

17. A rational number is a whole number.

18. Zero is a natural number.

19. A positive number is an integer.

Questions 20 and 21 refer to the following information.

Icy cold winter temperatures in Moscow, Russia, are commonplace.

20. If the temperature on January 1 increased from −14°F at 6 a.m. to 3°F at 3 p.m., how many degrees did the temperature increase?

21. After 3 p.m., the temperature started to decrease at a steady rate of about 0.5°F every hour. What was the temperature at 11 p.m.?

22. Evaluate $(-2)(-3)(6)$.

23. Evaluate $-\frac{1}{4}+\frac{3}{8}+\left(-\frac{5}{12}\right)$.

24. Convert 30% to a reduced fraction. Then, multiply that fraction by $\frac{25}{36}$. Write your answer in simplest form.

Questions 25 and 26 refer to the following information.

The Department of Transportation is planning to repave a busy $6\frac{3}{10}$-mile section of road and needs to complete the job in 3 days.

25. How many days would it take to finish if the average paver speed is about $1\frac{4}{5}$ miles per day?

26. How many miles per day would allow them to finish in 3 days?

27. Convert the fraction $\frac{1}{8}$ to a decimal and a percent.

28. Change $0.\overline{8}$ to a fraction.

29. What percent of 60 is 15?

For questions 30–32, read each expression and determine which value (C or D) is greater. Assume C and D are positive numbers.

30. 125% of C is D.

31. C is 14% of D.

32. 7% of C is D.

33. Earth is approximately 4.5 billion years old. Express this amount using scientific notation.

34. The blue whale is the heaviest living mammal ever known to have lived in the world and can weigh up to 4.0×10^5 pounds. The African elephant is the heaviest land mammal and can reach a weight of 1.4×10^4 pounds. How many times larger is the blue whale than the African elephant? Round your answer to the nearest whole number.

35. The speed of light is approximately 186×10^5 miles per second. If it takes about 5.04×10^2 seconds for the light from the Sun to reach Earth, about how many miles, to the nearest million, is the Sun from Earth?

Answers

1. 3, 4, 0, and −2 are integers.
2. $\frac{1}{2}$ and −5.8 are rational numbers and $\sqrt{6}$ is an irrational number.
3. 2, 5, and 7 are prime numbers.
4. 0
5. D
6. 25
7. A
8. 4^{13}
9. $\frac{1}{2^5}$
10. 3^6
11. D

12. approximately 7.3
13. Lucas multiplied first instead of dividing inside the parentheses (line 3). When deciding to multiply or divide in a numerical expression, always work from left to right and perform the operation that appears first. The correct answer is 0.
14. 3 and 7
15. $3\sqrt{6}$ inches
16. A. Always true, because integers are contained within the set of rational numbers.
17. B. Sometimes true, because rational numbers can be whole numbers, but they can represent any number that can be written as a fraction, like 3.2, which is not a whole number.
18. C. Never true; the set of natural numbers starts at the number 1.
19. B. Sometimes true; positive numbers can also be or contain decimals or fractions.
20. 17°
21. −1°F
22. 36
23. $-\frac{7}{24}$
24. $\frac{5}{24}$
25. 3.5 days
26. $2\frac{1}{10}$ miles per day
27. 0.125 and 12.5%
28. $\frac{8}{9}$
29. 25%
30. *D* is larger than *C*.
31. *D* is larger than *C*.
32. *C* is larger than *D*.
33. 4.5×10^9
34. $2.857 \times 10^1 = 28.57 \approx 29$
35. $9.3744 \times 10^9 = 9{,}374{,}400{,}000$ miles

Chapter 2

STRUCTURE IN ALGEBRAIC EXPRESSIONS

Chapter Check-In

- ❑ Set theory
- ❑ Structure of algebraic expressions
- ❑ Evaluating expressions

The topics covered in this chapter will help you to read, write, and interpret algebraic expressions so that you can manipulate and solve problems.

You will first be exploring and applying set theory to different quantitative situations. Set theory introduces the algebraic thinking necessary to understand the relationship among a collection of objects. Next, you will be working toward becoming algebraically proficient by applying mathematical operations to create equations that describe equivalent forms of numbers and variables. Using variables to write algebraic expressions is the foundation for a variety of Common Core Mathematics real-life algebraic scenarios.

Set Theory

Set theory and algebraic representation are two fundamental concepts that play an important role in Common Core Mathematics logic and reasoning. Set theory is used in almost every field of mathematics that deals with a collection of related objects.

Common Core Standard: Quantities

Reason quantitatively and define appropriate quantities for the purpose of descriptive modeling (N.Q.2). Choose a level of accuracy appropriate to limitations on measurement when reporting quantities (N.Q.3).

A **set** is used to describe a group or collection of distinct objects, such as the set of all oranges on a tree or the set of whole numbers less than 10. A set consisting of the numbers 1, 2, and 3 is written as {1, 2, 3}. It is common to use capital letters to name a set, such as T = {1, 2, 3}. The **elements** (or members) of each set are inside the braces separated by commas and do not repeat. To show that "3 is an element of the set {1, 2, 3}" symbolically, use the symbol ∈, which is read "is an element of" or "is a member of." Therefore, you could have written:

$$3 \in \{1, 2, 3\}$$

Special sets

A **subset** is a set contained within another set, or it can be the entire set itself. The set {1, 2} is a subset of the set {1, 2, 3}, and the set {1, 2, 3} is a subset of the set {1, 2, 3}. When the subset is missing some elements that are in the set it is being compared to, it is a proper subset. When the subset is the set itself, it is an improper subset. The symbol used to indicate "is a proper subset of" is ⊂. When there is the possibility of using an improper subset, the symbol used is ⊆. Therefore, $\{1, 2\} \subset \{1, 2, 3\}$ and $\{1, 2, 3\} \subseteq \{1, 2, 3\}$. The **universal set** is the general category set, or the set of all those elements under consideration. For example, if you are solving a problem to determine the number of students who took either Spanish or French in a high school, all the students in the high school are considered the universal set. The **empty set,** or **null set,** is the set with no elements or members. It is represented by the symbol ∅ or open braces { }. However, it is never represented by {∅}. The empty set is a subset of every set.

Types of sets

Finite sets have a *countable* number of elements. For example, {*a*, *b*, *c*, *d*, *e*} is a finite set because it contains only five elements. **Infinite sets** contain an *uncountable* number of elements. For example, the set of natural or counting numbers is N = {1, 2, 3, 4, 5, …}. The ellipsis mark (…) tells us that the set has an infinite number of elements; thus, it is an infinite set.

Universal sets can be either finite sets or infinite sets. The planets in the solar system make up a universal set that is finite. The set of real numbers is a universal set that is infinite.

Describing sets

Roster form is a method of naming a set by listing its members.

$\{1, 2, 3\}$ is a set consisting of the elements 1, 2, and 3.

$\{5, 10, 15, \ldots\}$ is a set consisting of the multiples of 5.

Set-builder notation is a method of naming a set by describing its elements. It uses two parts separated by either a vertical line or a colon. The first part names the variable, and the second part gives the properties that the elements within the set must satisfy. The table below illustrates various sets using the two different methods, along with the verbal representation of each set.

Set-Builder Notation	*Verbal Representation*	*Roster Form*
$\{x \mid x \text{ is a whole number, } x > 3\}$	The set of all values of x, such that x is a whole number greater than 3	$\{4, 5, 6, 7, 8, \ldots\}$
$\{x \mid x \text{ is a prime number, } x < 20\}$	The set of all real numbers, x, such that x is a prime number less than 20.	$\{2, 3, 5, 7, 11, 13, 17, 19\}$
$\{x \in \mathbb{Z} \mid -1 \le x \le 5\}$	All values of x that are elements of the set of integers, such that x is between -1 and 5, inclusive	$\{-1, 0, 1, 2, 3, 4, 5\}$

There are many ways to describe the set $\{1, 2, 3\}$ using set-builder notation. Three correct methods are as follows:

$$\{x \mid x \text{ is a natural number, } x < 4\}$$

$$\{x \mid x \text{ is a whole number, } 0 < x < 4\}$$

$$\{x \in \mathbb{N} \mid x < 4\}$$

An incorrect method would be $\{x \mid 0 < x < 4\}$, because this rule includes ALL numbers, such as decimals and fractions, between 0 and 4, not just the numbers 1, 2, and 3. This set of numbers cannot be represented as a list and is represented by using a number line graph, which is discussed in Chapter 7, "Inequalities, Graphing, and Absolute Value."

Note: Some sets of numbers use specific symbols for the names because they are used frequently in set theory. Some of the symbols you might see include:

$$\mathbb{R} = \text{Real Numbers}$$

$$\mathbb{Z} = \text{Integers}$$

$$\mathbb{Q} = \text{Rational Numbers}$$

$$\mathbb{N} = \text{Natural Numbers}$$

Comparing sets

Equal sets are sets that have the same members—{1, 2, 3} = {3, 2, 1}. **Equivalent sets** are sets that have the same number of members—{1, 2, 3} ~ {*a*, *b*, *c*}.

Venn diagrams (and **Euler circles**) are visual representations showing the relationship among elements to pictorially describe sets, as shown below. Visual references commonly appear in Common Core Mathematics problems.

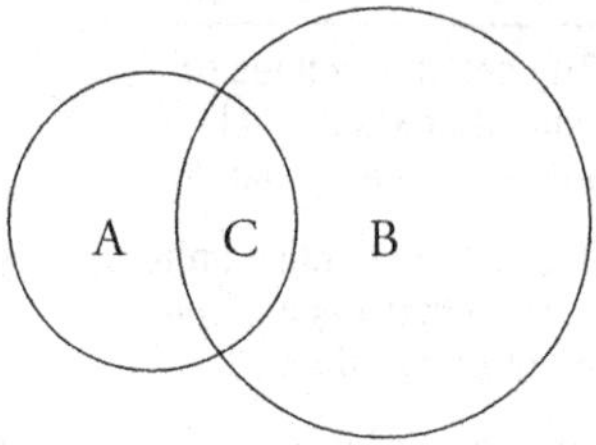

The A represents all the elements in the smaller circle; the B represents all the elements in the larger circle; and the C represents all the elements that are in both circles at the same time. Sometimes, you will see the circles inside a rectangle, which is the universal set, usually named with the capital letter U. ***Note:*** Venn diagrams can also be represented with ovals.

Complement of a set

The **complement of a set** describes all the elements that are in the universal set, but are not in the set being considered. The complement of set A is written as A′. For example, if the universal set, U = {the set of prime numbers} and A = {the set of prime numbers greater than 20}, then A′ = {2, 3, 5, 7, 11, 13, 17, 19}.

Operations with sets

The **union** of two or more sets is a set containing all the elements in those sets, but any duplicates are written only once. The symbol for finding the union of two sets is ∪.

Example 1: If A = {1, 2, 3} and B = {3, 4, 5}, what is A ∪ B?

$$A \cup B = \{1, 2, 3, 4, 5\}$$

The union of the set with members 1, 2, 3 together with the set with members 3, 4, 5 is the set with members 1, 2, 3, 4, 5.

The **intersection of sets** (two or more) is a set containing only the elements that are in both sets. The symbol for finding the intersection of two sets is ∩.

Example 2: If A = {1, 2, 3} and B = {3, 4, 5}, what is A ∩ B?

$$A \cap B = \{3\}$$

The intersection of the set with members 1, 2, 3 together with the set with members 3, 4, 5 is the set that has only the 3.

It is helpful to use Venn diagrams to illustrate these set relationships. The following Venn diagram displays the intersection and the union of sets A and B.

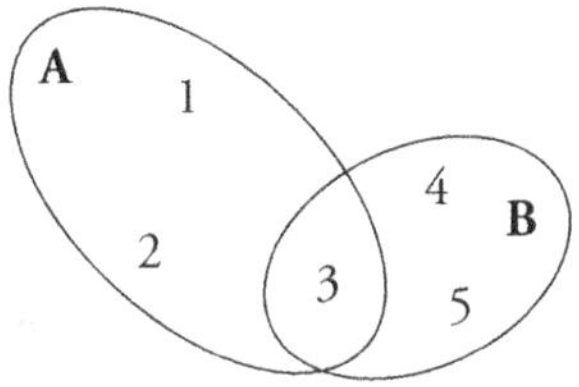

The union will be all the numbers represented in the diagram, {1, 2, 3, 4, 5}. The intersection would be where the two ovals overlap in the diagram, {3}.

Example 3: If A = {1, 2, 3} and C = {4, 5}, what is A ∩ C?

Since there are no members that are in both sets at the same time, then A ∩ C = ∅.

The intersection of the set with members 1, 2, 3 together with the set with members 4, 5 is the empty set, or null set. There are no members in both sets. These types of sets are called disjoint sets. **Disjoint sets** are sets that have no elements in common. The intersection of a set of disjoint sets is the empty set, ∅.

Structure of Algebraic Expressions

Common Core Mathematics expects that you accurately interpret algebraic expressions and persevere in solving problems. Use the strategies, tools, and exercises described in this section to help you look for the structure in algebraic expressions and apply reasoning skills to make sense of the operational relationships among variables.

Common Core Standard: Seeing the Structure in Expressions

Interpret parts of an expression that represent a quantity in terms of its context, such as terms, factors, and coefficients (A.SSE.1.a).

A **variable** is a symbol, usually a letter, used to represent an unknown number, and can also be used to denote any number in a specified set. Variables are used to translate verbal expressions into **algebraic expressions,** which are mathematical phrases (or sentences) that contain one or more variables.

For example, your friend is starting a business designing sandals and invests \$1,300 for equipment. The materials for each pair of sandals, p, cost \$20.

The expression that represents the cost in dollars of starting the sandal business is $20p + 1{,}300$.

Example 4: Write an algebraic expression for each word phrase.

Verbal Expression	*Algebraic Expression*
The sum of the number n and 7	$n + 7$ or $7 + n$
The number n diminished by 10	$n - 10$
Seven times the number n	$7n$
x divided by 4	$\frac{x}{4}$
Five more than the product of 2 and n	$2n + 5$ or $5 + 2n$
Seven less than the quotient of y and 4	$\frac{y}{4} - 7$
The number of hours in m minutes	$\frac{m}{60}$
The value of any number of dimes, d	$0.10d$
The cost to belong to a health club is a \$100 initiation fee plus \$75 per month	$100 + 75m$

Key words for operations

- **Key words denoting addition:**

sum	more than	gain
plus	greater than	grow
increase	rise	
enlarge	larger than	

- **Key words denoting subtraction:**

difference	smaller than	lower
minus	fewer than	diminish
lose	decrease	reduced
less than	drop	

- **Key words denoting multiplication:**

product	times	of
multiplied by	twice	

- **Key words denoting division:**

quotient	ratio	half
divided by		

Evaluating Expressions

Common Core Standard: Seeing the Structure in Expressions

Interpret parts of an expression such as terms, factors, and coefficients (A.SSE.1.a).

To evaluate an algebraic expression, take the following steps:

1. Substitute the given value(s) for the variable(s).
2. Simplify the expression, making sure to follow the rules for the order of operations.

Example 5: Evaluate each expression for the given values of the variables.

(a) $ab + c$ if $a = 5,\ b = 4,$ and $c = 3$

$$5(4) + 3 =$$
$$20 + 3 = 23$$

(b) $2x^2+3y+6$ if $x=2$ and $y=9$

$$2(2)^2+3(9)+6=$$
$$2(4)+27+6=$$
$$8+27+6=$$
$$35+6=41$$

(c) $-4p^2+5q-7$ if $p=-3$ and $q=-8$

$$-4(-3)^2+5(-8)-7=$$
$$-4(9)+5(-8)-7=$$
$$-36-40-7=-83$$

(d) $\frac{a+b}{c}+\frac{a}{b+c}$ if $a=3,\ b=-2,$ and $c=7$

$$\frac{3+(-2)}{7}+\frac{3}{(-2)+7}=$$
$$\frac{1}{7}+\frac{3}{5}=$$
$$\frac{5}{35}+\frac{21}{35}=\frac{26}{35}$$

(e) $5x^3y^2$ if $x=-2$ and $y=3$

$$5(-2)^3(3)^2=$$
$$5(-8)(9)=$$
$$-40(9)=-360$$

(f) The expression $2(wl + hl + hw)$ represents the surface area of a rectangular prism, where w is the width, h is the height, and l is the length. Betsy is painting a rectangular prism for a drama production, with a width of 6 inches, height of 2 inches, and length of 1 foot. How many square inches of paint will she need to cover the prism?

Since 1 foot = 12 inches, we substitute $h = 2$, $w = 6$, and $l = 12$ into the expression $2(wl + hl + hw)$.

Therefore, $2(6 \cdot 12 + 2 \cdot 12 + 2 \cdot 6) = 2(108) = 216\ \text{in}^2$.

Chapter Check-Out

Questions

Questions 1–4 refer to the following information.

Suppose $U = \{p \in \mathbb{Z} | -10 \leq p \leq 10\}$ is the universal set, $A = \{r \mid r$ is a composite number, $r < 10\}$, and $B = \{x \mid x$ is a whole number, $x \leq 4\}$.

1. Write sets A and B in roster form.
2. Find A ∩ B.
3. Find A ∪ B.
4. True or false: B ⊂ A.
5. Suppose U = {the oceans of the world} is the universal set and A = {Arctic Ocean}. What is A′?

Questions 6 and 7 refer to the following information.

H is the set of natural numbers between 3 and 10.

6. Using the information above, write in set-builder notation.
7. Using the information above, write in roster form.
8. If one set is a subset of another, can these two sets be disjoint? Explain your answer.
9. Write an algebraic expression for the word phrase, "Eight less than ten times the number *n*."
10. Write an algebraic expression for the word phrase, "The quotient of a number, *x*, and three that is increased by the product of five and another number, *y*."

Questions 11 and 12 refer to the following scenario.

Marta goes to the San Diego Zoo from 11:00 a.m. to 4:00 p.m. The cost of an admission ticket is $28.50 ,and the cost of metered parking is $2.25 per hour.

11. If *n* represents the number of tickets purchased, write an algebraic expression that gives the total cost of going to the zoo for the day.
12. If three of Marta's friends go with her to the zoo and they all come in one car, what is the total cost for the day?
13. Your friend writes $3n - 2$ to represent the phrase, "three times the difference of the number *n* and two." This algebraic expression is incorrect. Write the correct algebraic expression and explain the mistake made by your friend.

For question 14, select the correct answer from among the four choices.

14. At a local discount book store, the cost of each book is b dollars and the cost of each magazine is m dollars. If Henry buys 8 books and 4 magazines, what does the algebraic expression $8b + 4m$ represent?

A. the total number of books and magazines Henry owns
B. the total cost of books and magazines
C. the books and magazines
D. the number of books and the number of magazines Henry buys

15. Evaluate the expression $6x^2y + (2y)^3$ if $x = 2$ and $y = -4$.

16. Evaluate the expression $\frac{a-b}{c} - \frac{a+b}{2}$ if $a = 5$, $b = -1$, and $c = 3$.

17. If $h + m = 15$, what does $h + m + 15$ equal?

18. A student in your math class claims that $(x + y)^3$ is equivalent to $x^3 + y^3$ for all values of x and y. Is this claim correct? Justify your answer.

Question 19 refers to the following information.

The coldest layer of Earth's atmosphere is the mesosphere, which is about 50 to 80 kilometers above Earth's surface. The mesosphere protects Earth from most meteoroids since many meteoroids burn up in the mesosphere layer. The top of the mesosphere is the coldest part of the atmosphere, with temperatures that are as low as −90°C.

19. Use the information above and the following formula, $°F = \frac{9}{5}°C + 32$, to convert the Celsius temperature to Fahrenheit.

20. If $h = 4$, what is the value of the square root of h divided by the difference of twice h and six?

Answers

1. $A = \{4, 6, 8, 9\}$ and $B = \{0, 1, 2, 3, 4\}$
2. $A \cap B = \{4\}$
3. $A \cup B = \{0, 1, 2, 3, 4, 6, 8, 9\}$
4. False
5. A′ = {Antarctic (Southern) Ocean, Pacific Ocean, Atlantic Ocean, Indian Ocean}
6. $H = \{x \mid x \text{ is a natural number}, 3 < x < 10\}$ or $H = \{x \mid x \text{ is a natural number}, 4 \leq x \leq 9\}$
7. $H = \{4, 5, 6, 7, 8, 9\}$

8. If one set is a subset of another, the two sets can never be disjoint because the elements in one set are in the other set, which means the two sets have common elements.
9. $10n - 8$
10. $\frac{x}{3} + 5y$
11. $28.50n + 2.25(5)$
12. $125.25
13. The expression "three times the difference of the number n and 2" is written as $3(n - 2)$. Since we are multiplying 3 by the binomial $n - 2$, we need to use parentheses. The key word "difference" tells you to use parentheses to set off the subtraction.
14. B
15. -608
16. 0
17. 30
18. The claim that $(x + y)^3$ is equivalent to $x^3 + y^3$ is not correct. To prove, substitute various values for x and y, and you will find that this equivalency does not hold true for all values of x and y. For example, if $x = 2$ and $y = 3$,

$$(x+y)^3 \neq x^3 + y^3$$
$$(2+3)^3 \neq 2^3 + 3^3$$
$$(5)^3 \neq 8+27$$
$$125 \neq 35$$

19. -130°F
20. 1

Chapter 3
EQUATIONS WITH ONE VARIABLE

Chapter Check-In

- ❑ Properties of equality
- ❑ Creating and solving equations in one variable
- ❑ Literal equations
- ❑ Ratios and proportional relationships

Working with variables and solving equations often are considered the basis of algebra in Common Core Mathematics. When solving equations with one variable, the solution is the value of the variable that makes the equation true. In this chapter, you will be using inverse operations, properties of equality, and properties of operations to solve one-step and multi-step equations. This important algebraic concept will be extended further to solving equations for a given variable and rearranging formulas to solve for a specific quantity of interest.

Common Core Standard: Reasoning with Equations and Inequalities

Solve equations and inequalities in one variable, including equations with coefficients represented by letters (A.REI.3). Explain each step in solving a simple equation as following from the equality of numbers, starting from the assumption that the original equation has a solution. Construct a viable argument to justify a solution method (A.REI.1).

Equations

An **equation** is a mathematical statement, a relationship between numbers and/or variables that says two expressions are of equal value. An equation that contains numerical expressions is considered a **closed sentence** or **closed statement**. This means that the equation can be true *or false*. For example:

True: $3 + 4 = 7$

False: $6 - 8 = 2$

An equation that contains one or more variables is called an **open sentence**. This equation is neither true nor false, and the solution is unknown. For example, $2x - 4 = 8$ is an open sentence. An open sentence can have one solution, more than one solution, or no solution.

In the open sentence algebraic equation $x + 4 = 19$, some unknown number (represented by x) plus 4 equals 19. Using basic arithmetic, it is easy to see that $15 + 4 = 19$.

Variable (Unknown quantity) → $x + 4 = 19$ ← **Rational number** (Known quantity)

Equations that have the same solution are **equivalent equations.** For example, $2x - 4 = 8$ and $2x = 12$ are equivalent equations because they have the same solution, $x = 6$. The following properties of equality are used to justify the steps being taken when finding equivalent equations.

Properties of equality

The addition, subtraction, multiplication, and division properties of equality are used to solve equations. Let a, b, and c represent real numbers.

- **Reflexive property:** Any number equals itself.

  $a = a$

  Therefore, $4 = 4$.

- **Symmetric property:** If the first number is equal to the second number, then the second number is equal to the first number.

  If $a = b$, then $b = a$.

  Therefore, if $2 + 3 = 5$, then $5 = 2 + 3$.

- **Transitive property:** If the first number is equal to the second, and the second is equal to the third, then the first number is equal to the third.

  If $a = b$ and $b = c$, then $a = c$.

  Therefore, if $1 + 3 = 4$ and $4 = 2 + 2$, then $1 + 3 = 2 + 2$.

- **Addition property:** Adding the same number to each side of an equation gives an equivalent equation.

  If $a = b$, then $a + c = b + c$.

  Therefore, if $1 + 1 = 2$, then $1 + 1 + 3 = 2 + 3$.

- **Subtraction property:** Subtracting the same number from each side of an equation gives an equivalent equation.

  If $a = b$, then $a - c = b - c$.

  Therefore, if $4 + 2 = 6$, then $4 + 2 - 1 = 6 - 1$.

- **Multiplication property:** Multiplying by the same number on each side of an equation gives an equivalent equation.

  If $a = b$, then $ac = bc$.

  Therefore, if $1 = \frac{2}{2}$ and $4 = \frac{8}{2}$, then $1(4) = \left(\frac{2}{2}\right)\left(\frac{8}{2}\right)$.

- **Division property:** Dividing each side of an equation by the same number gives an equivalent equation.

  If $a = b$ and $c \neq 0$, then $\frac{a}{c} = \frac{b}{c}$.

  Therefore, if $2 + 1 = 3$, then $\frac{2+1}{4} = \frac{3}{4}$.

Creating and solving equations

Common Core Standard: Creating Equations

Create equations and inequalities in one variable and use them to solve problems. Include equations arising from linear and quadratic functions and simple rational and exponential functions (A.CED.1).

Represent constraints by equations or inequalities, and by systems of equations and/or inequalities, and interpret solutions as viable or non-viable options in a modeling context (A.CED.3). Rearrange formulas to highlight the quantity of interest, using the same reasoning as in solving equations (A.CED.4).

In general, solving an equation is a process of creating a series of equivalent equations until you isolate a variable with a coefficient of 1 on one side to find a solution. Equivalent equations are created by using the addition, subtraction, multiplication, and division properties of equality. Remember that an equation is like a balance scale with the equal sign (=) being the fulcrum, or center. Thus, if you apply the same operation to each side of the equation (say, add 5 to each side), the equation will still be balanced.

Equations will have either a unique solution, no solution, or infinitely many solutions. It is important to note that the solutions of an equation in one variable form a set of numbers, called a solution set. A **solution set** is the set of real numbers that will make the equation a true statement. When finding a solution, you will be manipulating the equation to keep each side balanced using inverse operations and properties of equality. **Inverse operations** are operations that undo each other. Remember that when you solve an equation, use inverse operations until the variable is on a side by itself (for addition, subtract; for multiplication, divide; and so forth). Examples 1 and 2 justify each step being taken in solving an equation.

Example 1: Solve for x: $x - 5 = 23$.

To find the solution of the equation $x - 5 = 23$, you must isolate x on one side of the equation; therefore, use the addition property of equality to undo subtraction by adding 5 to each side of the equation.

$$\begin{array}{r} x-5=23 \\ +5 \quad +5 \\ \hline x \quad\quad =28 \end{array}$$

In the same manner, you may subtract, multiply, or divide both sides of an equation by the same (nonzero) number, and the solution will not change. Sometimes you may have to use more than one step to solve an equation.

Example 2: Solve for x: $3x + 4 = 19$.

Use the subtraction property of equality to undo addition and subtract 4 from each side of the equation to isolate $3x$.

$$\begin{array}{r} 3x+4=19 \\ -4 \quad -4 \\ \hline 3x \quad\quad =15 \end{array}$$

Then use the division property of equality to undo multiplication and divide each side of the equation by 3 to find the solution.

$$\frac{3x}{3}=\frac{15}{3}$$
$$x=5$$

To check, substitute your solution into the original equation.

$$\begin{aligned}3x+4&=19\\3(5)+4&=19\\15+4&=19\\19&=19\checkmark\end{aligned}$$

Example 3: Solve for x: $\frac{x}{5}-4=2$.

Add 4 to each side of the equation.

$$\begin{aligned}\frac{x}{5}-4&=\ 2\\+4&\quad+4\\\hline\frac{x}{5}\qquad&=\ 6\end{aligned}$$

Multiply each side of the equation by 5 to find the solution.

$$5\left(\frac{x}{5}\right)=5(6)$$
$$x=30$$

Example 4: Solve for x: $\frac{3}{5}x-6=12$.

Add 6 to each side of the equation.

$$\begin{aligned}\frac{3}{5}x-6&=\ 12\\+6&\quad+6\\\hline\frac{3}{5}x\qquad&=\ 18\end{aligned}$$

Multiply each side of the equation by the reciprocal, $\frac{5}{3}$ (which is the same as dividing by $\frac{3}{5}$).

$$\frac{5}{3}\left(\frac{3}{5}x\right)=\frac{5}{3}\left(\frac{18}{1}\right)$$

$$x=\frac{5}{\cancel{3}_1}\left(\frac{\cancel{18}^6}{1}\right)$$

$$x=30$$

Example 5: Solve for x: $\frac{x-8}{2}=-3$.

Multiply each side of the equation by 2.

$$(2)\frac{x-8}{2}=-3(2)$$

$$x-8=-6$$

Add 8 to each side of the equation.

$$\begin{array}{r} x-8=-6 \\ +8\quad +8 \\ \hline x=2 \end{array}$$

Example 6: Solve for x: $16 = -2x + 4 - 4x$.

Combine like terms (x-terms) on the right side of the equation.

$$16 = -6x + 4$$

Subtract 4 from each side of the equation.

$$\begin{array}{l} 16=-6x+4 \\ -4\qquad\ -4 \\ \hline 12=-6x \end{array}$$

Divide each side of the equation by -6.

$$\frac{12}{-6}=\frac{\cancel{-6}x}{\cancel{-6}}$$

$$-2=x$$

Example 7: Solve for x: $5x = 2x - 6$.

Subtract $2x$ from each side of the equation.

$$\begin{array}{rr} 5x = & 2x - 6 \\ -2x & -2x \\ \hline 3x = & -6 \end{array}$$

Divide each side of the equation by 3.

$$\frac{3x}{3} = \frac{-6}{3}$$
$$x = -2$$

Example 8: Solve for x: $6x + 3 = 4x + 5$.

Subtract $4x$ from each side of the equation.

$$\begin{array}{rr} 6x + 3 = & 4x + 5 \\ -4x & -4x \\ \hline 2x + 3 = & 5 \end{array}$$

Subtract 3 from each side of the equation.

$$\begin{array}{rr} 2x + 3 = & 5 \\ -3 & -3 \\ \hline 2x = & 2 \end{array}$$

Divide each side of the equation by 2.

$$\frac{2x}{2} = \frac{2}{2}$$
$$x = 1$$

Identities and equations with no solution

Some equations are called **identity equations,** which means they are true for all values of the variable and the solution set is the set of all real numbers. Some equations have *no solution,* which means there are no values of the variable that make the equation true. Symbolically, you can represent no solution using { } or ∅.

Example 9: Solve for x: $3(2x + 1) = 3 + 6x$.

Use the distributive property.

$$6x + 3 = 3 + 6x$$

Subtract $6x$ from each side of the equation.

$$\begin{array}{r} 6x+3=3+6x \\ -6x \qquad -6x \\ \hline 3=3 \end{array}$$

Since $3 = 3$ is always true, this equation is an identity. This means that there are infinitely many solutions to this equation. Any value of x will make the equation true. For example, let's substitute $x = -1$ and $x = 5$ into the original equation to check.

Check:

$$\begin{array}{cc} x=-1 & x=5 \\ 3(2x+1)=3+6x & 3(2x+1)=3+6x \\ 3(2(-1)+1)=3+6(-1) & 3(2(5)+1)=3+6(5) \\ 3(-1)=3-6 & 3(11)=3+30 \\ -3=-3\checkmark & 33=33\checkmark \end{array}$$

Example 10: Solve for x: $3 + 5a = 5a - 4$.

Subtract $5a$ from each side of the equation.

$$\begin{array}{r} 3+5a=5a-4 \\ -5a \quad -5a \\ \hline 3 \quad = \quad -4 \end{array}$$

Since $3 \neq -4$, this equation has no solution. This means that you will not find one value of x that will make the equation true.

Literal equations

Literal equations are equations that contain two or more variables. When you work with literal equations, you will use the same properties of equality to isolate a specific variable of interest. For example, if you are writing the linear equation $x + 2y = 8$ in slope-intercept form (see Chapter 8, "Coordinate Geometry"), you will first need to solve the equation for y.

$$\begin{aligned} x+2y&=8 \\ 2y&=-x+8 \\ y&=-\frac{x}{2}+4 \end{aligned}$$

A *formula,* which is an equation that states a relationship among certain quantities, is a special type of literal equation used widely in mathematics and other related fields. The process of rewriting a literal equation is shown in the following examples.

Example 11: Solve for q: $qp - x = y$.

First add x to each side of the equation.

$$\begin{array}{rl} qp - x &= y \\ +x & +x \\ \hline qp &= y + x \end{array}$$

Then divide each side of the equation by p.

$$\frac{qp}{p} = \frac{y+x}{p}$$

$$q = \frac{y+x}{p}$$

Inverse operations were used to isolate q. To remove the $-x$, x was added to each side of the equation. Because the problem has q times p, each side of the equation was divided by p.

Example 12: Solve for y: $\frac{y}{x} = c$.

Multiply each side of the equation by x to isolate y.

$$x\left(\frac{y}{x}\right) = x(c)$$

$$y = xc$$

Example 13: Solve for x: $ax + b = cx + d$.

Subtract cx from each side of the equation.

$$\begin{array}{rl} ax \quad + b &= cx + d \\ -cx & -cx \\ \hline ax - cx + b &= d \end{array}$$

$$(a - c)x + b = d$$

By the distributive property, $ax - cx = (a - c)x$.

Subtract b from each side of the equation.

$$\begin{array}{r} (a-c)x+b=d \\ -b \quad -b \\ \hline (a-c)x=d-b \end{array}$$

Divide each side of the equation by $(a - c)$.

$$(a-c)x=d-b$$

$$\frac{(a-c)x}{(a-c)}=\frac{d-b}{a-c}$$

$$x=\frac{d-b}{a-c}$$

Example 14: Find the height of a trapezoid that has one base length of 6 inches, the other base length of 7 inches, and an area of 26 in^2. Use the formula $A=\frac{(b_1+b_2)h}{2}$.

Start by adjusting the formula so you're solving for h.

$$A=\frac{(b_1+b_2)h}{2}$$

Multiply each side of the equation by 2.

$$2(A)=\frac{(b_1+b_2)h}{2}(2)$$

$$2A=(b_1+b_2)h$$

Divide each side of the equation by $(b_1 + b_2)$.

$$\frac{2A}{(b_1+b_2)}=\frac{(b_1+b_2)h}{(b_1+b_2)}$$

$$\frac{2A}{(b_1+b_2)}=h$$

To find the height of a trapezoid with $b_1 = 6$, $b_2 = 7$, and $A = 26$ in^2, substitute these values into the following formula.

$$\frac{2A}{(b_1+b_2)} = h$$

$$\frac{2(26)}{(6+7)} = h$$

$$\frac{52}{13} = h$$

$$4 = h$$

The height of the trapezoid is 4 inches.

Ratios and Proportions

Common Core Standard: Reasoning with Equations and Inequalities

Explain each step in solving a simple equation as following from the equality of numbers, starting from the assumption that the original equation has a solution. Construct a viable argument to justify a solution method (A.REI.1). Solve linear equations and inequalities in one variable, including equations with coefficients represented by letters (A.REI.3).

Common Core Standard: Creating Equations

Rearrange formulas to highlight a quantity of interest, using the same reasoning as in solving equations (A.CED.4).

Ratios and proportions are not only used in arithmetic, but are also commonly used in the study of math and science to solve real-world and mathematical problems. Developing an understanding of ratios will help you analyze proportional relationships in algebraic equations. It is important to note that Common Core Mathematics focuses on using ratio and proportional reasoning to model and solve problems.

Ratios

A **ratio** is a method of comparing two or more quantities by division. Ratios are written as *a:b* or in working form, as a fraction.

$$a/b \text{ or } \frac{a}{b} \text{ is read "}a \text{ is to } b\text{"}$$

Notice that whatever comes after the "to" goes second, or in the denominator.

Equivalent ratios are ratios that have the same value, which means that when simplified, they all are the same ratio. For example, $\frac{5}{15}$, $\frac{2}{6}$, $\frac{3}{9}$, and $\frac{4}{12}$ are equivalent ratios because they all simplify to the same ratio, $\frac{1}{3}$, which equals $0.\overline{3}$. Two ratios that are equivalent are said to be proportional, or in proportion.

Proportions

Proportions are equations written as two equivalent ratios (fractions), in the form $\frac{a}{b} = \frac{c}{d}$, where $b \neq 0$ and $d \neq 0$. This is read as "*a* is to *b* as *c* is to *d*."

Example 15: Solve this proportion for *x*: *p* is to *q* as *x* is to *y*.

First the proportion may be rewritten.

$$\frac{p}{q} = \frac{x}{y}$$

Now, using the multiplication property of equality, multiply each side by *y*.

$$y\left(\frac{p}{q}\right) = y\left(\frac{x}{y}\right)$$

$$\frac{yp}{q} = x$$

Example 16: Solve this proportion for *t*: *s* is to *t* as *r* is to *q*.

Rewrite.

$$\frac{s}{t} = \frac{r}{q}$$

Multiply each side of the equation by the least common denominator, *tq*, and simplify by canceling common factors.

$$(\not{t})(q)\frac{s}{\not{t}} = (t)(\not{q})\frac{r}{\not{q}}$$

$$qs = tr$$

Now, using the division property of equality, divide each side by r.

$$\frac{qs}{r} = \frac{t\cancel{r}}{\cancel{r}}$$

$$\frac{qs}{r} = t$$

Note: The result is the same if you use the **cross products property**. Using the proportion $\frac{s}{t} = \frac{r}{q}$, the products qs and tr are called *cross products.*

Cross products of a proportion are equal. Therefore, $qs = tr$.

Divide each side of the equation by r.

$$\frac{qs}{r} = \frac{tr}{r}$$

$$\frac{qs}{r} = t \text{ or } t = \frac{qs}{r}$$

Solving proportions for value

If two ratios are proportional and one quantity is unknown, you can either multiply both sides of the equation by the least common denominator or use the cross products property to find the unknown quantity.

Least Common Denominator	Cross Products Property
$\frac{10}{5} = \frac{x}{12}$	$\frac{10}{5} = \frac{x}{12}$
$\overset{12}{\cancel{(60)}}\frac{10}{\underset{1}{\cancel{5}}} = \frac{x}{\underset{1}{\cancel{12}}}\overset{5}{\cancel{(60)}}$	$5x = 10(12)$
$\frac{120}{5} = \frac{5x}{5}$	$5x = 120$
$24 = x$	$x = 24$

Examples 17 and 18 use the cross products property to solve each proportion.

Example 17: Solve for x: $\frac{4}{x} = \frac{2}{5}$.

Use the cross products property.

$$(4)(5) = 2x$$

$$20 = 2x$$

Divide each side of the equation by 2.

$$\frac{20}{2}=\frac{2x}{2}$$
$$10=x$$

Example 18: Solve for x: $\frac{2x-4}{3x+6}=\frac{2}{5}$.

Use the cross products property.

$$5(2x-4)=2(3x+6)$$

Use the distributive property.

$$10x-20=6x+12$$

Subtract $6x$ from each side of the equation.

$$\begin{array}{rcr} 10x-20 & = & 6x+12 \\ -6x & & -6x \\ \hline 4x-20 & = & 12 \end{array}$$

Add 20 to each side of the equation.

$$\begin{array}{rcr} 4x-20 & = & 12 \\ +20 & & +20 \\ \hline 4x & = & 32 \end{array}$$

Divide each side of the equation by 4.

$$\frac{4x}{4}=\frac{32}{4}$$
$$x=8$$

Chapter Check-Out

Questions

1. True or false: If $a=b$ and $b=c$, then $a=c$.
2. If $x-322=76$, what does $x-233$ equal?

For question 3, select the correct answer from among the four choices.

3. Eddy asked Rebecca to write an equation that has a solution of -4. Which of the following equations is possible?

   A. $-4=x-4$
   B. $-4x=4$
   C. $\frac{x}{2}-7=-5$
   D. $3x=5x+8$

4. Write down each algebraic property used in finding the solution to the following equation.

$$\frac{5}{3}x+2(x-3)=-10$$

(a) $\frac{5}{3}x+2x-6=-10$ ____________

(b) $\frac{11}{3}x-6=-10$ ____________

(c) $\frac{11}{3}x=-4$ ____________

(d) $x=-\frac{12}{11}$ ____________

5. Describe and correct the error in finding the solution of the equation below.

$$\frac{-4+x}{3}=-15$$

$$\frac{x}{3}=-11$$

$$x=-33$$

6. Solve for x: $7x + 3 = 5x + 7$.

Question 7 refers to the following information and equation.

The equation below is true when $x = 2$.

$$12 - x = 2(13 - 4x)$$

7. Using the information above, what value of x will make the following equation true?

$$12-\frac{x}{2}=2\left(13-4\left(\frac{x}{2}\right)\right)$$

8. Solve for m: $mn - r = q$.

For question 9, select the correct answer from among the four choices.

9. Megan wrote the equation $x = 4y - 1$. If the value of y is increased by 2, the value of x will

A. decrease by 1
B. decrease by 2
C. increase by 4
D. increase by 8

10. Solve the proportion for y: m is to n as y is to z.

For question 11, select the correct answer from among the four choices.

11. Which proportion does not have the same solution as $\frac{6}{x} = \frac{3}{5}$?

A. $\frac{4}{x} = \frac{2}{5}$

B. $\frac{5}{6} = \frac{10}{2x-6}$

C. $\frac{3}{10} = \frac{12}{4x}$

D. $\frac{3x+2}{4x} = \frac{4}{5}$

Questions 12 and 13 refer to the following information.

The volume of a pyramid can be calculated using the formula $V = \frac{1}{3}Bh$, where B is the area of the base and h is the height of the pyramid.

12. Write an expression that can be used to represent the height, h.

13. Find the height of a square pyramid that is 3 feet long on each side and has a volume of 23 cubic feet.

14. Ted solves the equation $4(2x - 3) = 8$ by using the distributive property, but Emily finds the solution to the equation by using the division property of equality. Show that both methods will give the same solution.

For questions 15–17, select one of the following three answer choices to determine whether the statement is:

A. always true
B. sometimes true
C. never true

Explain your answers. Let a and b represent real numbers.

15. The equation in the form $bx + 1 = bx$ has one solution.

16. An equation in one variable has more than one solution.

17. The proportion $\frac{ax}{2} = \frac{bx}{2}$ has no solution.

Answers

1. True
2. 165
3. D
4. **(a)** distributive property; **(b)** add like terms; **(c)** addition property of equality; **(d)** multiplication property of equality
5. The error made was adding 4 to both sides of the equation first. You must either multiply both sides by 3 first, or write the left side as a difference of two fractions and then add $-\frac{4}{3}$ to both sides.

Method 1

$$\frac{-4+x}{3} = -15$$

$$\not{3}\left(\frac{-4+x}{\not{3}}\right) = 3 \cdot -15$$

$$-4+x = -45$$

$$x = -41$$

Method 2

$$\frac{-4+x}{3} = -15$$

$$-\frac{4}{3} + \frac{x}{3} = -15$$

$$\frac{x}{3} = -15 + \frac{4}{3}$$

$$\not{3}\left(\frac{x}{\not{3}}\right) = \not{3}\left(-\frac{41}{\not{3}}\right)$$

$$x = -41$$

6. $x = 2$
7. $x = 4$

8. $m = \frac{q+r}{n}$

9. D

10. $y = \frac{mz}{n}$

11. B

12. $h = \frac{3V}{B}$

13. $h = 7\frac{2}{3}$ feet

14.

Ted's Method	Emily's Method
$4(2x-3)=8$	$\frac{4(2x-3)}{4} = \frac{8}{4}$
$8x-12=8$	$2x-3=2$
$+12\ +12$	$+3\ +3$
$\frac{8x}{8} = \frac{20}{8}$	$\frac{2x}{2} = \frac{5}{2}$
$x=2.5$	$x=2.5$

15. Never true. The equation $bx + 1 = bx$ has no solution.

16. Sometimes true. An equation in one variable can either have one solution, no solution, or infinitely many solutions.

17. Sometimes true. If $a = b$, the equation is an identity and has infinitely many solutions. If $a \neq b$, the equation has no solution.

Chapter 4

EQUATIONS WITH TWO VARIABLES

Chapter Check-In

- ❑ Solving systems of equations
- ❑ Elimination method
- ❑ Substitution method
- ❑ Graphing method

If you have two different equations with the same two variables in each, you can solve for both variables.

Solving Systems of Equations (Simultaneous Equations)

Common Core Standard: Creating Equations

Create equations in two or more variables to represent relationships between quantities; graph equations in the coordinate plane with labels and scales (A.CED.2). Represent constraints by equations or inequalities, and by systems of equations and/or inequalities, and interpret solutions as viable or non-viable options in a modeling context (A.CED.3).

Common Core Standard: Reasoning with Equations and Inequalities

Prove that, given a system of two equations in two variables, replacing one equation by the sum of that equation and a multiple of the other produces a system with the same solutions (A.REI.5). Solve systems of linear equations exactly and approximately, focusing on pairs of linear equations in two variables (A.REI.6).

A **system of equations,** or a **system,** is a collection of two or more equations with the same set of variables. A system that contains two linear equations with two variables (x and y) is called a *system of linear equations.* A single solution of a system of linear equations is an ordered pair of numbers (x, y) that makes each equation in the system true. The solution can be found using three common methods: elimination, substitution, and graphing.

Since systems of equations can be used to solve many real-world problems containing two or more variables (linear or non-linear), this topic plays an important role in Common Core Mathematics and its practical applications to areas in physics, engineering, computer science, economics, chemistry, earth science, and many more.

Elimination method

The **elimination method** is also known as the addition/subtraction method. To solve a system using elimination, you will be adding or subtracting the equations to eliminate one of the variables. This method is best to use if the equations are in standard form ($Ax + By = C$), and the coefficients of one of the variables can easily be made the same or opposite.

To use the elimination method, do the following:

1. Multiply one or both equations, if needed, by a number to make the coefficients of one variable the same or exactly the opposite in each equation.
2. Add or subtract the two equations to eliminate one variable.
3. Solve for the remaining variable.
4. Solve for the eliminated variable by substituting the value of the variable found into one of the original equations.

Note: A **coefficient** is a number in front of a term that contains a variable(s); it multiplies the variable. In the expression $5x$, the 5 is "in front" of the x. (For more on coefficients, refer to Chapter 5, "Monomials, Polynomials, and Factoring.")

Example 1: Solve the system using elimination.

$$x + y = 7$$
$$x - y = 3$$

Since the coefficients of each of the y terms are opposites, adding the equations eliminates the y terms.

$$\begin{aligned} x+y&=7 \\ x-y&=3 \\ \hline 2x \quad &=10 \\ \frac{2x}{2}&=\frac{10}{2} \\ x&=5 \end{aligned}$$

Now, substituting 5 for x in the first equation gives the following:

$$\begin{aligned} 5+y&=\ \ 7 \\ -5 \quad &\ \ -5 \\ \hline y&=\ \ 2 \end{aligned}$$

The solution is (5, 2).

By substituting each x with a 5 and each y with a 2 in the original equations, you can see that each equation will be made true.

Example 2: Solve the system using elimination.

$$\begin{aligned} 3x+3y&=24 \\ 2x+y&=13 \end{aligned}$$

First, multiply the bottom equation by 3. Now, the coefficient of y is 3 in each equation.

$$\begin{aligned} 3x+3y&=24 \qquad & 3x+3y&=24 \\ 3(2x)+3(y)&=3(13) \qquad & 6x+3y&=39 \end{aligned}$$

Now, the equations can be subtracted, eliminating the y terms.

$$\begin{aligned} 3x+\ \ 3y\ &=\ \ 24 \\ -6x+(-3y)\ &\ \ -39 \\ \hline -3x \qquad &=-15 \\ \frac{-3x}{-3}&=\frac{-15}{-3} \\ x&=5 \end{aligned}$$

Substitute $x = 5$ in one of the original equations to solve for y.

$$\begin{aligned} 2x + y &= 13 \\ 2(5) + y &= 13 \\ 10 + y &= 13 \\ -10 \quad & \quad -10 \\ \hline y &= 3 \end{aligned}$$

The solution is (5, 3).

Of course, if the number in front of a variable is already the same in each equation, you do not have to change either equation. Simply add or subtract.

To check the solution, substitute each x in each equation with 5 and substitute each y in each equation with 3.

$$\begin{aligned} 3x + 3y &= 24 \\ 3(5) + 3(3) &= 24 \\ 15 + 9 &= 24 \\ 24 &= 24 \checkmark \end{aligned} \qquad \begin{aligned} 2x + y &= 13 \\ 2(5) + (3) &= 13 \\ 10 + 3 &= 13 \\ 13 &= 13 \checkmark \end{aligned}$$

In Example 1 and Example 2, a unique answer existed for x and y that made each equation true at the same time. In other words, there is a single (x, y) value that satisfies both equations. In some situations, a system may have no solution or infinitely many solutions. Be aware of these when you use the elimination method.

Example 3: Solve the system using elimination.

$$\begin{aligned} 3a + 4b &= 2 \\ 6a + 8b &= 4 \end{aligned}$$

Multiply the top equation by 2. Notice what happens.

$$\begin{aligned} 2(3a) + 2(4b) &= 2(2) \\ 6a + 8b &= 4 \end{aligned} \qquad \begin{aligned} 6a + 8b &= 4 \\ 6a + 8b &= 4 \end{aligned}$$

Now if you were to subtract one equation from the other, the result is $0 = 0$. This statement is *always true,* which means that there are infinitely many (x, y) values that will satisfy both equations.

When this occurs, the system of equations has infinitely many solutions. Any value of a and b that makes one of the equations true also makes the other equation true. For example, if $a = -6$ and $b = 5$, then both equations are made true.

$$3(-6) + 4(5) = 2 \text{ and } 6(-6) + 8(5) = 4$$

What we have here is really only one equation written in two different ways. In this case, the second equation is actually the first equation multiplied by 2.

Example 4: Solve the system using elimination.

$$3x+4y=5$$
$$6x+8y=9$$

Multiply the top equation by 2. Notice what happens.

$$2(3x)+2(4y)=2(5) \qquad 6x+8y=10$$
$$6x+8y=9 \qquad 6x+8y=9$$

Now if you were to subtract the bottom equation from the top equation, the result would be $0 = 1$. This statement is *never true,* which means there are no (x, y) values that will satisfy both equations. When this occurs, the system of equations has no solution.

In Examples 1–4, only one equation was multiplied by a number to make the coefficients the same or opposite. Sometimes each equation must be multiplied by different numbers for the coefficients to be the same or opposite.

Example 5: Solve the system using elimination.

$$3x+4y=5$$
$$5x-6y=2$$

Notice that there is no simple number you can multiply either equation with for the coefficients of x or y to become the same or opposites. In this case, do the following:

1. Select a variable to eliminate.
2. Use the two coefficients of this variable. Find the least common multiple of this value as the desired coefficient to be in front of each variable.
3. Determine what value each equation needs to be multiplied by to obtain this value and multiply the equation by that number.

Suppose you want to eliminate x. The least common multiple of 3 and 5 is 15, so the coefficient of x needs to be changed to 15. The first equation must be multiplied by 5 in order to get $15x$. The second equation must be multiplied by 3 in order to get $15x$.

$$5(3x)+5(4y)=5(5) \qquad 15x+20y=25$$
$$3(5x)-3(6y)=3(2) \qquad 15x-18y=6$$

Now subtract the second equation from the first equation to get the following:

$$38y=19$$
$$y=\frac{19}{38} \text{ or } y=\frac{1}{2}$$

At this point, you can either substitute $\frac{1}{2}$ for y and solve for x (method 1 that follows), or start with the original two equations and eliminate y in order to solve for x (method 2 that follows).

Method 1. Using the top equation, substitute $\frac{1}{2}$ for y and solve for x.

$$3x+4y=5$$
$$3x+4\left(\frac{1}{2}\right)=5$$
$$3x+2=5$$
$$3x=3$$
$$x=1$$

Method 2. Eliminate y and solve for x.

The least common multiple of 4 and 6 is 12. Multiply the top equation by 3 and the bottom equation by 2 to get the coefficient 12.

$$3(3x)+3(4y)=3(5) \qquad 9x+12y=15$$
$$2(5x)-2(6y)=2(2) \qquad 10x-12y=4$$

Now add the two equations to eliminate y.

$$19x = 19$$
$$x = 1$$

The solution is $\left(1, \frac{1}{2}\right)$.

Substitution method

Sometimes a system is more easily solved by the **substitution method.** This method involves substituting one equation into another. This method is best to use when one of the equations is already solved for a specific variable or when you can easily solve one of the equations for a specific variable.

Example 6: Solve the system using substitution.

$$x = y + 8$$
$$x + 3y = 48$$

From the first equation, substitute $(y + 8)$ for x in the second equation.

$$(y + 8) + 3y = 48$$

Now solve for y. Simplify by combining like terms (y terms).

$$\begin{array}{r} 4y + 8 = 48 \\ -8 \quad -8 \\ \hline 4y \quad = 40 \end{array}$$
$$\frac{4y}{4} = \frac{40}{4}$$
$$y = 10$$

Now substitute y's value, 10, into one of the original equations.

$$x = y + 8$$
$$x = 10 + 8$$
$$x = 18$$

The solution is (18, 10).

Check the solution.

$$\begin{aligned} x &= y+8 \\ 18 &= 10+8 \\ 18 &= 18\checkmark \end{aligned} \qquad \begin{aligned} x+3y &= 48 \\ 18+3(10) &= 48 \\ 18+30 &= 48 \\ 48 &= 48\checkmark \end{aligned}$$

Example 7: Solve the system using substitution.

$$\begin{aligned} 5x+6y &= 14 \\ y-4x &= -17 \end{aligned}$$

First, find a variable that has either a 1 or -1 as a coefficient. Solve for that variable in terms of the other variable.

Then proceed as in Example 6.

In this example, the bottom equation has a coefficient of 1 for y.

Solve for y in terms of x.

$$\begin{aligned} y-4x &= -17 \\ y &= 4x-17 \end{aligned}$$

Substitute $4x - 17$ for y in the top equation and then solve for x.

$$\begin{aligned} 5x+6y &= 14 \\ 5x+6(4x-17) &= 14 \\ 5x+24x-102 &= 14 \\ 29x-102 &= 14 \\ 29x &= 116 \\ x &= \frac{116}{29} = 4 \end{aligned}$$

Substitute 4 for x in the second equation, $y - 4x = -17$, and solve for y.

$$\begin{aligned} y-4x &= -17 \\ y-4(4) &= -17 \\ y-16 &= -17 \\ y &= -1 \end{aligned}$$

The solution is (4, −1).

Check the solution:

$$\begin{aligned} 5x+6y &= 14 \\ 5(4)+6(-1) &= 14 \\ 20-6 &= 14 \\ 14 &= 14 \checkmark \end{aligned} \qquad \begin{aligned} y-4x &= -17 \\ -1-4(4) &= -17 \\ -1-16 &= -17 \\ -17 &= -17 \checkmark \end{aligned}$$

Graphing method

Another method of solving a system is by graphing each equation on a coordinate graph. The **graphing method** gives a visual representation of the solution, which is the point of intersection (x, y) of the two straight lines. If you are unfamiliar with coordinate graphing, carefully review Chapter 8, "Coordinate Geometry," before attempting this method.

Example 8: Solve the system by graphing.

$$\begin{aligned} x &= 4+y \\ x-3y &= 4 \end{aligned}$$

First, find three values for x and y that satisfy each equation. (Although only two points are necessary to determine a straight line, finding a third point is a good way of checking.) Following are tables of x and y values:

$x = 4 + y$

x	y
4	0
2	−2
5	1

$x - 3y = 4$

x	y
1	−1
4	0
7	1

Now graph the two lines on the coordinate plane, as shown in the following figure.

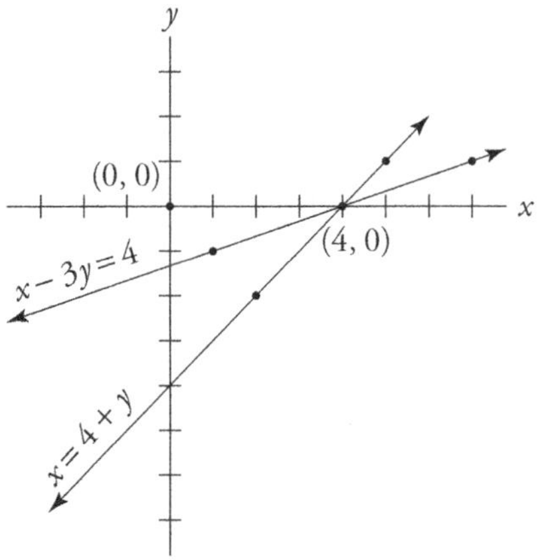

The point where the two lines intersect (4, 0) is the solution of the system.

Example 9: Solve the system by graphing.

$$3x + 4y = 2$$
$$6x + 8y = 4$$

Find three values for x and y that satisfy each equation. Following are the tables of x and y values.

$3x + 4y = 2$

x	y
0	$\frac{1}{2}$
2	−1
4	$-\frac{5}{2}$

$6x + 8y = 4$

x	y
0	$\frac{1}{2}$
2	−1
4	$-\frac{5}{2}$

Now graph the two lines on the coordinate plane, as shown in the following figure.

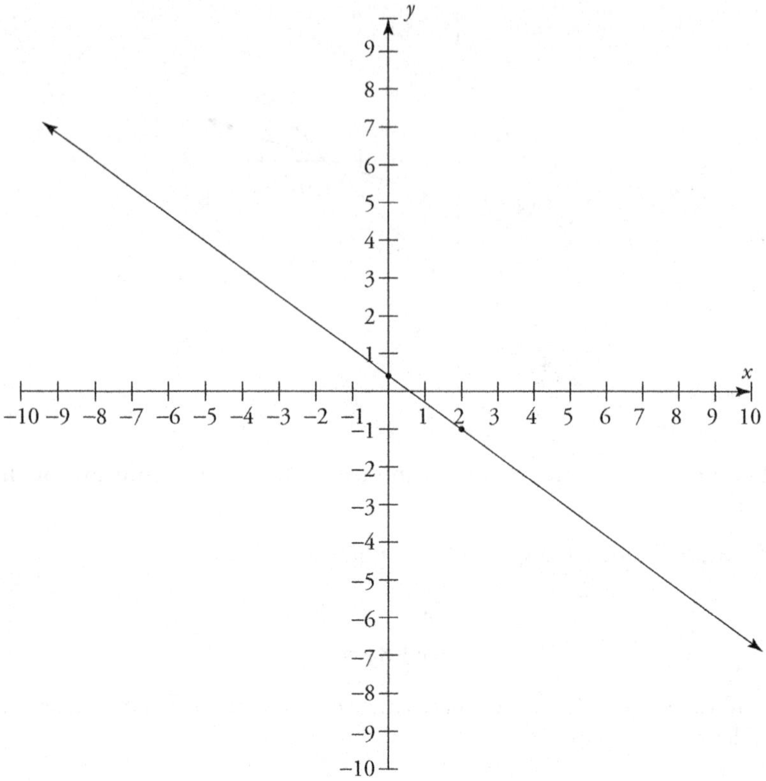

Notice that the same points satisfy each equation. These equations represent the same line.

Therefore, the solution is not a unique point. Since any point on the line is a solution to the system, there are infinitely many solutions. Notice that systems with infinitely many solutions have the same slope and the same y-intercept. This is like Example 3 when it was solved using the elimination method.

Example 10: Solve the system of equations by graphing.

$$3x + 4y = 4$$
$$6x + 8y = 16$$

Find three values for x and y that satisfy each equation. See the following tables of x and y values:

$3x + 4y = 4$

x	y
0	1
2	$-\frac{1}{2}$
4	-2

$6x + 8y = 16$

x	y
0	2
2	$\frac{1}{2}$
4	-1

Now graph the two lines on the coordinate plane, as shown in the following figure.

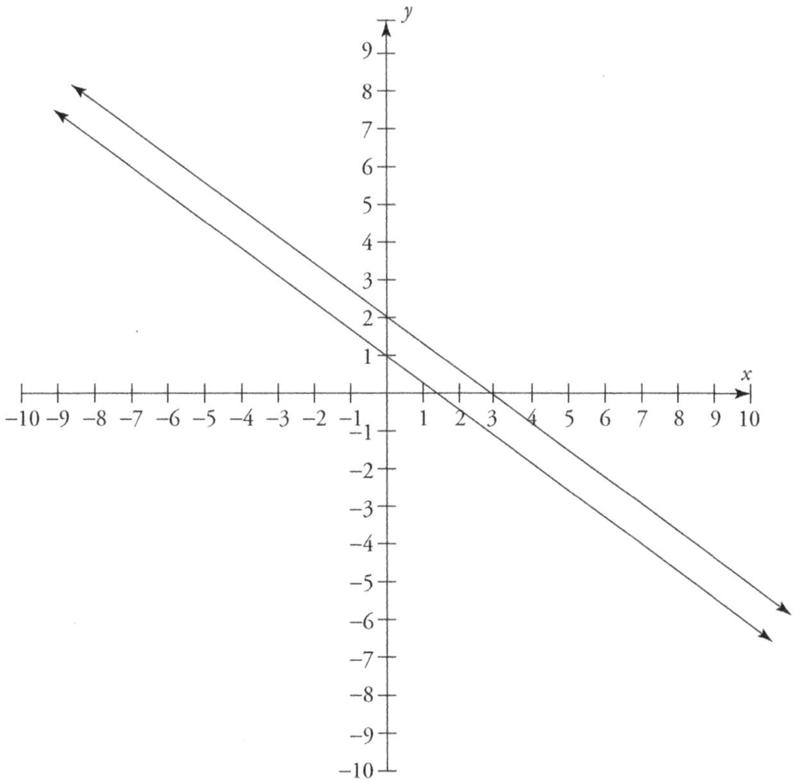

The two lines are parallel. They will never meet. Therefore, there is no solution for this system of equations. Notice that systems with no solution have the same slope and different y-intercepts. This is like Example 4, which was solved using the elimination method.

When solving some linear systems using the graphing method, the solution may be inexact (with non-integer coordinates). If you are not using graphing technology to find a solution, solving systems by graphing is not the desired method to use. Although you can use the graphing method to approximate solutions, the substitution or elimination methods will give an exact solution.

Example 11: Estimate the solution to the system by graphing, then solve the system algebraically.

$$y = 2x - 3$$
$$2x + 4y = 10$$

$y = 2x - 3$

x	**y**
0	−3
4	5
−1	−5

$2x + 4y = 10$

x	**y**
5	0
1	2
−1	3

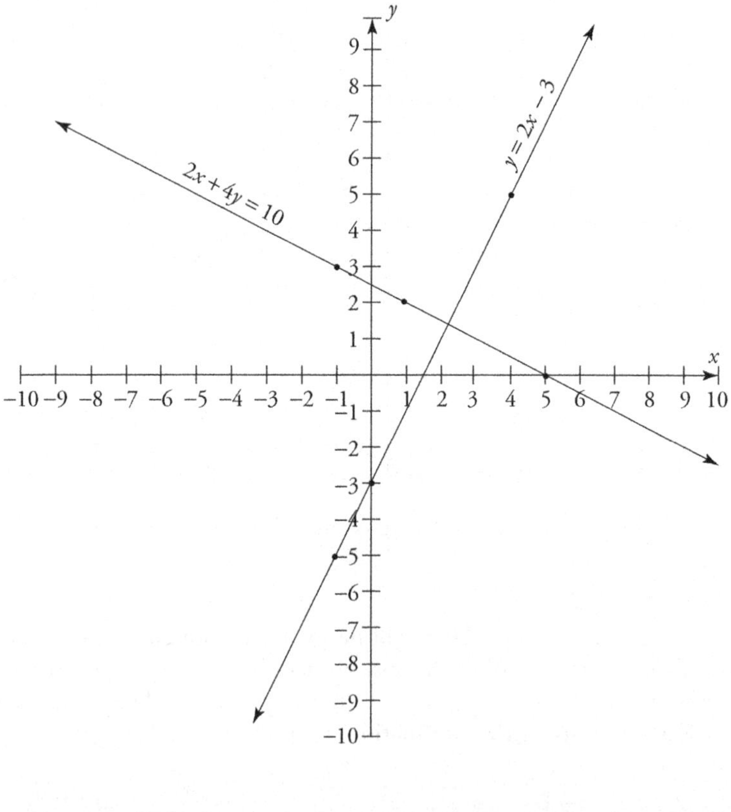

Estimate: (2.1, 1.5)

In Example 11, solving using substitution is the most efficient method since the top equation is already solved for y.

Substitute $(2x - 3)$ for y in the second equation and then solve for x.

$$\begin{aligned} 2x+4(2x-3) &= 10 \\ 2x+8x-12 &= 10 \\ 10x-12 &= 10 \\ 10x &= 22 \\ x &= \frac{22}{10} = 2.2 \end{aligned}$$

Substitute 2.2 for x in the first equation, $y = 2x - 3$.

$$\begin{aligned} y &= 2x-3 \\ y &= 2(2.2)-3 \\ y &= 1.4 \end{aligned}$$

The solution is (2.2, 1.4).

Chapter Check-Out

Questions

1. Use the elimination method to solve for x and y.

$$\begin{aligned} 8x+2y &= 7 \\ 3x-4y &= 5 \end{aligned}$$

2. Use the substitution method to solve for a and b.

$$\begin{aligned} a &= b+1 \\ a+2b &= 7 \end{aligned}$$

3. Use the graphing method to solve for x and y.

$$\begin{aligned} 2x-3y &= 0 \\ x+y &= 5 \end{aligned}$$

Question 4 refers to the following information and equation.

A student used the elimination method to find the solution to the system below.

$$\begin{array}{lll} 4x+3y=10 \rightarrow & 3(4x+3y=10) \rightarrow & 12x+9y=30 \\ 6x+2y=15 \rightarrow & -2(6x+2y=15) \rightarrow & \underline{-12x-4y=15} \\ & & 5y=45 \\ & & y=9,\ x=-11 \\ & & \text{Solution: } (-11,\ 9) \end{array}$$

4. **(a)** Describe and correct the error, and **(b)** find the solution to the system.
5. Use any method to solve the system of equations.

$$3x+y=3$$
$$6x+2y=12$$

6. After solving a system of equations, Rebecca shows the answer is $4 = 4$.

   (a) What does this tell Rebecca about the system of equations?

   (b) If Rebecca graphs the two linear equations, what would she notice about the lines?
7. Suppose you find that a system of linear equations is true when $x = 3$ and $y = -5$. What conclusion can you make about the graphs of the linear equations?
8. Find a, such that the graphs of $y = ax - 5$ and $y = -3x + 4$ have no solution.

   Hint: These equations are in slope-intercept form, $y = mx + b$, where m is the slope and b is the y-intercept.
9. Zhang graphs the system $y = -2x + 5$ and $y = -3x - 1$ on the next page and finds that there is no solution. Describe and correct Zhang's error.

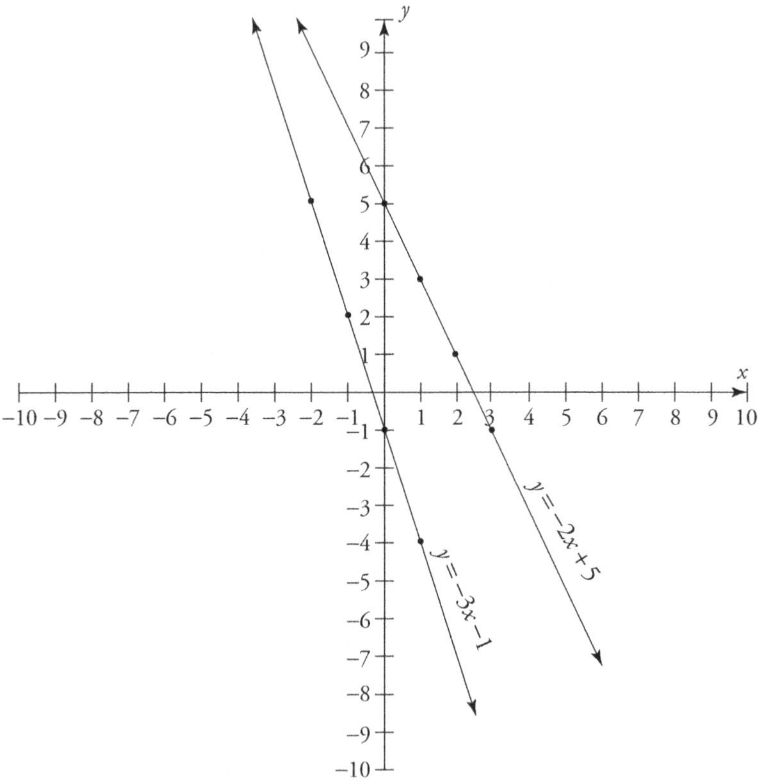

10. You find the solution to the system $\begin{matrix} 4x - 3y = 10 \\ x + 2y = -3 \end{matrix}$ to be (1, −2), but your friend concludes that the solution is (4, 2). Explain how you can check to see which ordered pair is the correct solution.

Answers

1. $\left(1, -\frac{1}{2}\right)$
2. (3, 2)
3. (3, 2)

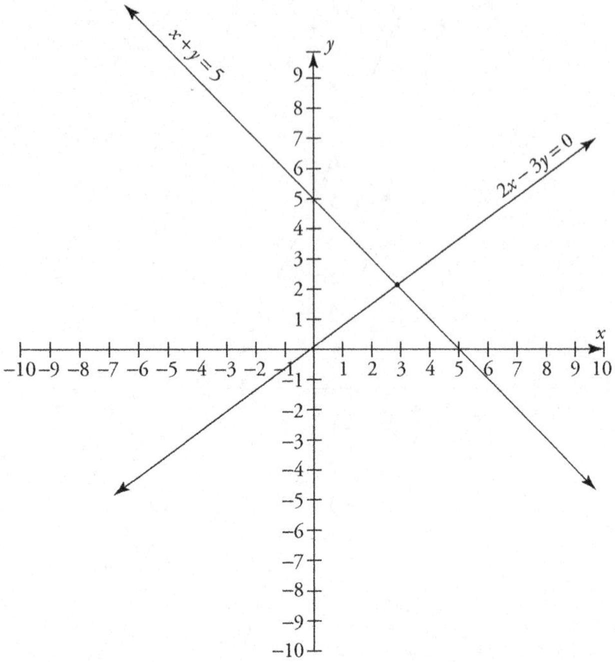

4. (a) When the student multiplied the bottom equation by -2, the student forgot to multiply by the constant. The bottom equation, when multiplied by -2, is $-12x - 4y = -30$.

   (b) The correct solution is (2.5, 0).

5. No solution

6. (a) The answer $4 = 4$ tells Rebecca that there are infinitely many solutions for the system of equations.

   (b) If she were to graph the system, the lines would coincide (they are exactly the same line).

7. The graphs of the linear equations will intersect at the point $(3, -5)$ on the coordinate plane.

8. $a = -3$

9. Zhang did not extend the lines long enough to see the intersection point. The lines will intersect at $(-6, 17)$.

10. To find the correct solution, check each ordered pair by substituting the (x, y) values into both equations. When substituting $(1, -2)$ into both equations, they are both true. When substituting $(4, 2)$ into both equations, the bottom equation is not true. So, $(1, -2)$ is the correct solution to the system.

Chapter 5

MONOMIALS, POLYNOMIALS, AND FACTORING

Chapter Check-In

- ❑ Monomials: Operations with monomials and working with negative exponents
- ❑ Polynomials: Operations with polynomials
- ❑ Factoring

One of the basic skills used in Common Core Mathematics algebra is the ability to work with monomials and polynomials. Factoring polynomial expressions is another important basic skill, but before you can factor, you must review the basics.

Monomials

Common Core Standard: Arithmetic with Polynomials and Rational Expressions

Understand that polynomials form a system analogous to the integers. Namely, they are closed under the operations of addition, subtraction, and multiplication. Add, subtract, and multiply polynomials (A.APR.1). Rewrite simple rational expressions in different forms; write $\frac{a(x)}{b(x)}$ in the form $\frac{q(x)+r(x)}{b(x)}$, where $a(x)$, $b(x)$, $q(x)$, and $r(x)$ are polynomials with the degree of $r(x)$ less than the degree of $b(x)$, using inspection, long division, or, for the more complicated examples, a computer algebra system (A.APR.6).

(Continued)

Common Core Standard: Seeing Structure in Expressions

Interpret parts of an expression, such as terms, factors, and coefficients (A.SSE.1.a). Use the structure of an expression to identify ways to rewrite it (A.SSE.2).

Common Core Standard: The Real Number System

Explain how the definition of rational exponents follows from extending the properties of integer exponents to those values, allowing for a notation for radicals in terms of rational exponents (N.RN.1).

A **monomial** is an algebraic expression that consists of only one term. This term can be a real number, a variable, or a product of real numbers and variables with whole-number exponents. For example, $9x$, $4a^2$, -2, x, and $3mpx^2$ are all monomials. However, $\frac{9}{x}$ and $4x\sqrt[3]{y}$ are not monomials because the exponents of the variables are not whole numbers. $\frac{9}{x}$ can be written as $9x^{-1}$ and $4x\sqrt[3]{y}$ can be written as $4xy^{\frac{1}{3}}$. The number in front of the variable is called the **numerical coefficient.** In $9x$, 9 is the coefficient. Note that if there is no number in front of the variable, the coefficient is assumed to be 1. The coefficient of x is 1. The **degree of a monomial** is the sum of the exponents of its variables. For example, the degree of x is 1 and the degree of $5x^2y^2$ is 4. If the monomial is a constant, like 6, the degree is 0, since 6 can be written as $6x^0$.

Adding and subtracting monomials

To add or subtract monomials, follow the same rules as with signed numbers (see Chapter 1, "Foundations for Understanding Algebra"), provided that the monomials are like terms. Monomials that are **like terms** have the same variable factors and can be combined by addition or subtraction. Notice that you add or subtract only the coefficients and leave the variables the same. For example, $-9x^2$ and $7x^2$ are like terms; when combined, they will simplify to $-2x^2$. However, $-9x$ and $7x^2$ are not like terms and cannot be combined.

Example 1: Simplify each expression.

(a)

$$\begin{array}{r} 15x^2yz \\ -18x^2yz \\ \hline -3x^2yz \end{array}$$

(b) $3x + 2x = 5x$

(c)
$$\begin{array}{r} 9y \\ -3y \\ \hline 6y \end{array}$$

(d)
$$\begin{aligned} 17q+8q-3q-(-4q)&= \\ 22q-(-4q)&= \\ 22q+4q&=26q \end{aligned}$$

Remember that the rules for signed numbers apply to monomials as well.

Multiplying monomials

Reminder: When multiplying or dividing powers with the same base, the properties of exponents also apply with algebraic expressions (see Chapter 1, "Foundations for Understanding Algebra").

$5 \cdot 5 = 5^2$ and $x \cdot x = x^2$
Similarly, $a \cdot a \cdot a \cdot b \cdot b = a^3b^2$.

To multiply monomials, multiply the coefficients (if any) and add the exponents of the variables with the same bases.

Example 2: Simplify each expression.

(a) $(x^3)(x^4) = x^{3+4} = x^7$

(b) $(x^2y)(x^3y^2) = (x^2x^3)(yy^2) = x^{2+3}y^{1+2} = x^5y^3$

(c) $(6k^5)(5k^2) = (6 \times 5)(k^5k^2) = 30k^{5+2} = 30k^7$ (multiply coefficients)

(d) $-4(m^2n)(-3m^4n^3) = [(-4)(-3)](m^2m^4)(nn^3) = 12m^{2+4}n^{1+3} = 12m^6n^4$ (multiply coefficients)

(e) $(c^2)(c^3)(c^4) = c^{2+3+4} = c^9$

(f) $(3a^2b^3c)(b^2c^2d) = 3(a^2)(b^3b^2)(cc^2)(d) = 3a^2b^{3+2}c^{1+2}d = 3a^2b^5c^3d$

Note that in example (d), the product of -4 and -3 is 12, the product of m^2 and m^4 is m^6, and the product of n and n^3 is n^4 because any monomial with no exponent indicated is assumed to have an exponent of 1.

When a monomial is being raised to a power, raise the coefficient (if any) to the power and multiply the exponent of each base by the exponent of the power to which it is being raised.

Example 3: Simplify each expression.

(a) $(a^7)^3 = a^{21}$

(b) $(x^3y^2)^4 = x^{12}y^8$

(c) $(2x^2y^3)^3 = (2)^3x^6y^9 = 8x^6y^9$

Dividing monomials

To divide monomials, divide or reduce the coefficients (if any) and subtract the exponent of the divisor from the exponent of the dividend of the variables with the same base.

Example 4: Simplify each expression.

(a) $\dfrac{y^{15}}{y^4} = y^{11}$ or $y^{15} \div y^4 = y^{11}$

(b) $\dfrac{x^5y^2}{x^3y} = x^2y$

(c) $\dfrac{36a^4b^6}{-9ab} = -4a^3b^5$ (divide the coefficients)

(d) $\dfrac{fg^{15}}{g^3} = fg^{12}$

(e) $\dfrac{x^5}{x^8} = \dfrac{1}{x^3}$ (may also be expressed as x^{-3})

(f) $\dfrac{-3(xy)(xy^2)}{xy}$

You can simplify the numerators first.

$$\frac{-3(xy)(xy^2)}{xy} = \frac{-3x^2y^3}{xy} = -3xy^2$$

Or, because the numerator is all multiplication, you can reduce.

$$\frac{-3\left(\overset{1}{\cancel{x}}\,\overset{1}{\cancel{y}}\right)(xy^2)}{\underset{1}{\cancel{x}}\,\underset{1}{\cancel{y}}} = -3xy^2$$

Working with negative exponents

Remember, if the exponent is negative, such as x^{-3}, take the reciprocal of the base and change the exponent from negative to positive. That is, the variable and exponent may be dropped under the number 1 in a fraction to remove the negative sign as follows.

$$x^{-3} = \frac{1}{x^3}$$

An algebraic expression with exponents is in its simplest form when powers with a variable base contain only positive exponents. When a negative exponent is in the numerator, move the base to the denominator and the exponent becomes positive. Similarly, when a negative exponent is in the denominator, move the base to the numerator and the exponent becomes positive.

Example 5: Simplify each expression.

(a) $a^{-2}b = \frac{b}{a^2}$

(b) $\frac{a^{-3}}{b^4} = \frac{1}{a^3b^4}$

(c) $\left(a^2b^{-3}\right)\left(a^{-1}b^4\right) = ab$

$$\begin{bmatrix} a^2 \cdot a^{-1} = a \\ b^{-3} \cdot b^4 = b \end{bmatrix}$$

(d) $\frac{-3}{6y^{-3}} = \frac{-1y^3}{2}$

Polynomials

Common Core Standard: Arithmetic with Polynomials and Rational Expressions

Use polynomial identities to solve problems (A.APR). Prove polynomial identities and use them to describe relationships. Understand the relationship between zeros and factors of polynomials (A.APR.4).

Common Core Standard: Seeing Structure in Expressions

Interpret parts of an expression, such as terms, factors, and coefficients (A.SSE.1.a). Use the structure of an expression to identify ways to rewrite it (A.SSE.2).

A **polynomial** is an algebraic expression that consists of one or more monomials. For example, $x + y$, $y^2 - x^2$, and $x^2 + 3x + 5y^2$ are all polynomials. A **binomial** is a polynomial that consists of exactly two monomials. For example, $x + y$ is a binomial. A **trinomial** is a polynomial that consists of exactly three monomials. For example, $y^2 + 9y + 8$ is a trinomial. The **degree of a polynomial** is the highest degree of its terms in the simplified expression. For example, the degree of $2x^3 + x^2 + 5x$ is 3.

Polynomials in one variable are usually arranged in one of two ways: ascending order or descending order. In **ascending order,** the degrees of the terms increase for each succeeding term. For example, $x + x^2 + x^3$ or $5x + 2x^2 - 3x^3 + x^5$ are arranged in ascending order. In **descending order,** the degrees of the terms decrease for each succeeding term. For example, $x^3 + x^2 + x$ or $2x^4 + 3x^2 + 7x$ are arranged in descending order. Descending order is more commonly used, and the polynomial is said to be in standard form when written in descending order.

Adding and subtracting polynomials

To add or subtract polynomials, arrange like terms in columns and then add or subtract the coefficients of the like terms. (Simply add or subtract like terms when rearrangement is not necessary.) The polynomial will be in simplest form when all the like terms are combined.

Example 6: Simplify each expression.

(a)
$$\begin{array}{r} a^2 + ab + b^2 \\ \underline{3a^2 + 4ab - 2b^2} \\ 4a^2 + 5ab - b^2 \end{array}$$

(b) $(5y - 3x) + (9y + 4x) = (5y + 9y) + (-3x + 4x) = 14y + x$ or $x + 14y$

(c)
$$\begin{array}{rcr} a^2 + b^2 & & a^2 + b^2 \\ \underline{-(2a^2 - b^2)} & \rightarrow & \underline{-2a^2 + b^2} \\ & & -a^2 + 2b^2 \end{array}$$

(d) $$\begin{aligned}(3cd-6mt)-(2cd-4mt) &= (3cd-6mt)+(-2cd+4mt)\\ &= (3cd-2cd)+(-6mt+4mt)\\ &= cd-2mt\end{aligned}$$

(e) $3a^2bc+2ab^2c+4a^2bc+5ab^2c=$

$$\begin{array}{r} 3a^2bc+2ab^2c \\ +4a^2bc+5ab^2c \\ \hline 7a^2bc+7ab^2c \end{array}$$

or

$$\begin{aligned}3a^2bc+2ab^2c+4a^2bc+5ab^2c &= (3\text{a}^2bc+4a^2bc)+(2ab^2c+5ab^2c)\\ &= 7a^2bc+7ab^2c\end{aligned}$$

It is important to note that polynomials are closed under addition and subtraction, which means that when you add or subtract polynomials, the sum or difference is a polynomial.

Multiplying polynomials

To multiply polynomials, use the distributive property to multiply each term in one polynomial by each term in the other polynomial. Then simplify if necessary.

Example 7: Multiply: $(3x + a)(2x - 2a)$.

$(3x+a)(2x-2a)=3x(2x-2a)+a(2x-2a)$	Distribute each term of the first binomial to the second binomial.
$=6x^2-6ax+a(2x-2a)$	Distribute $3x$.
$=6x^2-6ax+2ax-2a^2$	Distribute a.
$=6x^2-4ax-2a^2$	Combine like terms.

Or you may want to use the **F.O.I.L. method** with binomials. F.O.I.L. stands for First terms, Outside terms, Inside terms, Last terms. Then simplify if necessary.

Example 8: Multiply: $(3x + a)(2x - 2a)$.

Multiply the first terms from each quantity.

$$\overset{\downarrow \text{ first } \downarrow}{(3x+a)(2x-2a)} = 6x^2 \_\_\_\_\_\_\_\_$$

Next, multiply the outside terms.

$$\overset{\downarrow \text{ outside } \downarrow}{(3x+a)(2x-2a)} = 6x^2 - 6ax \_\_\_\_\_\_\_\_$$

Then multiply the inside terms.

$$\overset{\downarrow \text{ inside } \downarrow}{(3x+a)(2x-2a)} = 6x^2 - 6ax + 2ax \_\_\_\_\_\_\_\_$$

Finally, multiply the last terms.

$$\overset{\downarrow \text{ last } \downarrow}{(3x+a)(2x-2a)} = 6x^2 - 6ax + 2ax - 2a^2$$

Now simplify.

$$6x^2 - 6ax + 2ax - 2a^2 = 6x^2 - 4ax - 2a^2$$

Example 9: Multiply: $(x + y)(x + y + z)$.

$$\begin{array}{r} x+y+z \\ \times \qquad\qquad x+y \\ \hline xy+y^2+yz \\ +x^2+xz+xy \qquad\quad \\ \hline x^2+xz+2xy+y^2+yz \end{array}$$

This operation also can be done using the distributive property.

$$\begin{aligned}(x+y)(x+y+z) &= x(x+y+z)+y(x+y+z) \\ &= x^2+xy+xz+xy+y^2+yz \\ &= x^2+2xy+xz+yz+y^2\end{aligned}$$

It is important to note that just like addition and subtraction, polynomials are also closed under multiplication. The product of two polynomials will produce another polynomial.

Dividing polynomials by monomials

To divide a polynomial by a monomial, just divide each term of the polynomial in the numerator by the monomial in the denominator.

Example 10: Divide.

(a) $(6x^2+2x)\div(2x)=$

$$\frac{6x^2+2x}{2x}=$$

$$\frac{6x^2}{2x}+\frac{2x}{2x}=3x+1$$

(b) $(16a^7-12a^5)\div(4a^2)=$

$$\frac{16a^7-12a^5}{4a^2}=$$

$$\frac{16a^7}{4a^2}-\frac{12a^5}{4a^2}=4a^5-3a^3$$

Dividing polynomials by polynomials

To divide a polynomial by a polynomial, make sure both are in standard form; then use long division. (***Remember:*** Divide by the first term, multiply, subtract, bring down.) Similar to long division of real numbers, your answer will be written as

$$\text{quotient}+\frac{\text{remainder}}{\text{divisor}}$$

If the remainder is zero, then the quotient is a factor of the polynomial.

Example 11: Divide $4a^2 + 18a + 8$ by $a + 4$.

First divide $4a^2$ by a.

$$\begin{array}{r} 4a \\ a+4\overline{)4a^2+18a+8} \end{array}$$

Note: Align terms with the same degree ($4a$ is placed above $18a$).

Now multiply $4a$ times $(a + 4)$.

$$\begin{array}{r} 4a \\ a+4\overline{)4a^2+18a+8} \\ 4a^2+16a \end{array}$$

Now subtract.

$$\begin{array}{r} 4a \\ a+4\overline{\big)\ 4a^2+18a+8} \\ \underline{-(4a^2+16a)} \\ 2a \end{array}$$

Now bring down the 8.

$$\begin{array}{r} 4a \\ a+4\overline{\big)\ 4a^2+18a+8} \\ \underline{-\left(4a^2+16a\right)} \\ 2a+8 \end{array}$$

Now divide 2*a* by *a*.

$$\begin{array}{r} 4a+2 \\ a+4\overline{\big)\ 4a^2+18a+8} \\ \underline{-\left(4a^2+16a\right)} \\ 2a+8 \end{array}$$

Now multiply 2 times $(a + 4)$.

$$\begin{array}{r} 4a+2 \\ a+4\overline{\big)\ 4a^2+18a+8} \\ \underline{-\left(4a^2+16a\right)} \\ 2a+8 \\ 2a+8 \end{array}$$

Now subtract.

$$\begin{array}{r} 4a+2 \\ a+4\overline{\big)\ 4a^2+18a+8} \\ \underline{-\left(4a^2+16a\right)} \\ 2a+8 \\ \underline{-(2a+8)} \\ 0 \end{array}$$

$$\begin{array}{r} 4a+2 \\ a+4\overline{)\;4a^2+18a+8} \\ \underline{-(4a^2+16a)} \\ 2a+8 \\ \underline{-(2a+8)} \\ 0 \end{array} \qquad \text{similar to} \qquad \begin{array}{r} 23 \\ 53\overline{)\;1219} \\ \underline{-106} \\ 159 \\ \underline{-159} \\ 0 \end{array}$$

Note: $4a + 2$ and $a + 4$ are factors of $4a^2 + 18a + 8$. So, $(4a + 2)(a + 4) = 4a^2 + 18a + 8$.

Example 12: Divide.

(a) $(3x^2 + 4x + 1) \div (x + 1)$

$$\begin{array}{r} 3x+1 \\ x+1\overline{)\;3x^2+4x+1} \\ \underline{-(3x^2+3x)} \\ x+1 \\ \underline{-(x+1)} \\ 0 \end{array}$$

(b) $(2x + 1 + x^2) \div (x + 1) =$

First change to descending order: $x^2 + 2x + 1$. Then divide

$$\begin{array}{r} x+1 \\ x+1\overline{)\;x^2+2x+1} \\ \underline{-(x^2+\;x)} \\ x+1 \\ \underline{-(x+1)} \\ 0 \end{array}$$

(c) $(m^3 - m) \div (m + 1) =$

Note: If the dividend has a missing term, add the missing term with a coefficient of zero when dividing.

$$
\begin{array}{r}
m^2 - m \\
m+1 \overline{\big)\; m^3 + 0m^2 - m} \\
\underline{-(m^3 + \;\; m^2)} \\
-m^2 - m \\
\underline{-(-m^2 - m)} \\
0
\end{array}
$$

(d) $(10a^2 - 29a - 21) \div (2a - 7) =$

$$
\begin{array}{r}
5a + 3 \\
2a-7 \overline{\big)\; 10a^2 - 29a - 21} \\
\underline{-(10a^2 - 35a)} \\
6a - 21 \\
\underline{-(6a - 21)} \\
0
\end{array}
$$

(e) $(x^2 + 2x + 4) \div (x + 1) =$

$$
\begin{array}{r}
x + 1 \\
x+1 \overline{\big)\; x^2 + 2x + 4} \\
\underline{-(x^2 + \;\; x)} \\
x + 4 \\
\underline{-(x + 1)} \\
3
\end{array}
\quad \text{(with remainder 3)}
$$

This answer can be written as $x + 1 + \dfrac{3}{x+1}$.

Factoring

Common Core Standard: Seeing Structure in Expressions

Interpret parts of an expression, such as terms, factors, and coefficients (A.SSE.1.a). Use the structure of an expression to identify ways to rewrite it (A.SSE.2).

In algebra, factoring polynomials is used to simplify expressions, simplify fractions, and solve equations. To **factor** means to find two or more quantities whose product equals the original quantity. When factoring a polynomial, we write the expression as the product of its factors. For example, when we factor the polynomial $x^2 + 3x + 2$, we get $(x + 1)(x + 2)$. This means that the product of the factors $(x + 1)$ and $(x + 2)$ equals the original polynomial, $x^2 + 3x + 2$. It is important to note that factoring a polynomial is the reverse process of multiplying polynomials, so the distributive property is applied throughout this section.

Factoring a monomial from a polynomial

To factor a monomial from a polynomial, first find the greatest common factor (GCF) of its terms, and then divide the original polynomial by this factor to obtain the second factor (or apply the distributive property to factor out the GCF). The second factor will be a polynomial.

Example 13: Factor each polynomial.

(a) $5x^2 + 4x = x(5x + 4)$

(b) $2y^3 - 6y = 2y(y^2 - 3)$

(c) $x^5 - 4x^3 + x^2 = x^2(x^3 - 4x + 1)$

When the common monomial factor is the last term, 1 is used as a placeholder in the second factor.

Factoring the difference between two squares

To factor the difference between two squares, first find the square root of the first term and of the second term, and then express your answer as a product of two binomials that are, respectively, the sum and the difference of those square roots.

Example 14: Factor each expression.

(a) $x^2 - 144 = (x + 12)(x - 12)$ ***Note:*** $x^2 + 144$ is not factorable.

(b) $a^2 - b^2 = (a + b)(a - b)$

(c) $9y^2 - 1 = (3y + 1)(3y - 1)$

Factoring trinomials of the form $ax^2 + bx + c$

To factor trinomials of the form $ax^2 + bx + c$, first check to see whether you can factor out a GCF. Then if $a = 1$ (that is, the first term is simply x^2), use two sets of parentheses and factor the first term. Place these factors in the left sides of the two sets of parentheses. For example, $(x \quad)(x \quad)$.

Next, factor the last term and place the factors in the right sides of the two sets of parentheses.

To decide on the signs of the numbers, do the following.

- If the sign of the last term is *negative*, find two factors of the last term (one will be a positive number and the other a negative number) whose difference is the coefficient of the middle term and give the larger of these two numbers the sign of the middle term and the other number the *opposite* sign.
- If the sign of the last term is *positive*, find two factors of the last term (both will be positive or both will be negative) whose sum is the coefficient of the middle term and give both factors the sign of the middle term.

Example 15: Factor $x^2 - 3x - 10$.

First check to see whether you can factor out a GCF. Because this is not possible, use two sets of parentheses and factor the first term as follows: $(x \quad)(x \quad)$. Next, find two factors of 10 whose difference is -3. The factors 2 and -5 have a product of -10 and a difference of -3, leaving

$$(x - 5)(x + 2)$$

Multiply means (inner terms) and extremes (outer terms) to check.

$$\overbrace{\downarrow (x-5)(x+2) \downarrow}^{\text{outer } 2x} \qquad 2x - 5x = -3x \text{ (which is the middle term)}$$

$$\underbrace{\uparrow\ \uparrow}_{-5x \text{ inner}}$$

To completely check, multiply the binomial factors together.

$$\begin{array}{r} x-5 \\ \times \quad x+2 \\ \hline +2x-10 \\ +x^2-5x \quad \\ \hline x^2-3x-10 \end{array}$$

Example 16: Factor $x^2 + 8x + 15$.

$$(x + 3)(x + 5)$$

Notice that $3 \times 5 = 15$ and $3 + 5 = 8$, the coefficient of the middle term. Also note that the signs of both factors are positive, the sign of the middle term. To check,

$$\overset{\text{outer}}{\overset{5x}{\overline{\downarrow \qquad \downarrow}}}$$

$$(x+3)(x+5) \qquad 5x+3x=8x \text{ (which is the middle term)}$$

$$\underset{\text{inner}}{\underset{3x}{\underline{\uparrow\ \uparrow}}}$$

Example 17: Factor $x^2 - 5x - 14$.

$$(x - 7)(x + 2)$$

Notice that $7 \times 2 = 14$ and $7 - 2 = 5$, the coefficient of the middle term. Also note that the sign of the larger factor, 7, is negative, while the other factor, 2, is positive. To check,

$$\overset{\text{outer}}{\overset{2x}{\overline{\downarrow \qquad \downarrow}}}$$

$$(x-7)(x+2) \qquad 2x-7x=-5x \text{ (which is the middle term)}$$

$$\underset{\text{inner}}{\underset{-7x}{\underline{\uparrow\ \uparrow}}}$$

If, however, the first term has a coefficient other than 1 (for example, $4x^2 + 5x + 1$), then additional trial and error will be necessary.

Example 18: Factor $4x^2 + 5x + 1$.

$(2x + \quad)(2x + \quad)$ might work for the first term. But when 1s are used as factors to get the last term, $(2x + 1)(2x + 1)$, the middle term comes out as $4x$ instead of $5x$.

$$\overset{\text{outer}}{\overset{2x}{\overbrace{(2x+1)(2x+1)}}} \qquad 2x + 2x = 4x$$

(inner: $2x$)

Therefore, try $(4x + \quad)(x + \quad)$. Now using 1s as factors to get the last terms gives $(4x + 1)(x + 1)$. Checking for the middle term,

$$\overset{\text{outer}}{\overset{4x}{\overbrace{(4x+1)(x+1)}}} \qquad 4x + x = 5x$$

(inner: x)

Therefore, $4x^2 + 5x + 1 = (4x + 1)(x + 1)$.

Example 19: Factor $4a^2 + 6a + 2$.

Factoring out a 2 leaves

$$2(2a^2 + 3a + 1)$$

Now factor as usual, giving

$$2(2a + 1)(a + 1)$$

To check,

outer
$2a$
$\downarrow \qquad \downarrow$

$(2a+1)(a+1)$ $\quad 2a + a = 3a$ (the middle term after 2 was factored out)

$\uparrow\uparrow$
a
inner

Example 20: Factor $5x^3 + 6x^2 + x$.

Factoring out an x leaves

$$x(5x^2 + 6x + 1)$$

Now factor as usual, giving

$$x(5x + 1)(x + 1)$$

To check,

outer
$5x$
$\downarrow \qquad \downarrow$

$(5x+1)(x+1)$ $\quad 5x + x = 6x$ (the middle term after x was factored out)

$\uparrow\uparrow$
x
inner

Example 21: Factor $5 + 7b + 2b^2$ (a slight twist).

$$(5 + 2b)(1 + b)$$

To check,

$$\overbrace{(5+2b)(1+b)}^{\text{outer } 5b} \qquad 5b+2b=7b \text{ (the middle term)}$$

$$\underbrace{\uparrow\ \uparrow}_{\substack{2b\\ \text{inner}}}$$

Note that $(5 + b)(1 + 2b)$ is incorrect because it gives the wrong middle term.

Example 22: Factor $x^2 + 2xy + y^2$.

$$(x + y)(x + y)$$

To check,

$$\overbrace{(x+y)(x+y)}^{\text{outer } xy} \qquad xy+xy=2xy \text{ (the middle term)}$$

$$\underbrace{\uparrow\ \uparrow}_{\substack{xy\\ \text{inner}}}$$

Example 23: Factor $3x^2 - 48$.

Factoring out a 3 leaves

$$3(x^2 - 16)$$

But $x^2 - 16$ is the difference between two squares and can be further factored into $(x + 4)(x - 4)$. Therefore, when completely factored,

$$3x^2 - 48 = 3(x + 4)(x - 4)$$

Factoring by grouping

Some polynomials have binomial, trinomial, and other polynomial factors.

Example 24: Factor $x + 2 + xy + 2y$.

Since there is no monomial factor, you should attempt to rearrange the terms and look for binomial factors.

$$x + 2 + xy + 2y = x + xy + 2 + 2y$$

Grouping gives

$$(x + xy) + (2 + 2y)$$

Now factoring gives

$$x(1 + y) + 2(1 + y)$$

Factoring out the common binomial $(1 + y)$ gives

$$(x + 2)(1 + y)$$

You could rearrange them differently, but you would still come up with the same factors.

You can also factor by grouping when working with trinomials of the form $ax^2 + bx + c$ by writing bx as the sum of two monomials. This method can be more efficient when factoring trinomials where $a > 1$.

Examples 18 through 21 took you through the process of factoring using trial and error. Examples 25 and 26 below use algebraic reasoning to take you through the process of factoring by grouping.

Example 25: Factor $2x^2 + 3x - 9$.

Step 1: Find the product of the leading coefficient, 2, and the constant, −9.

$$2(-9) = -18$$

Step 2: Find two factors of −18 that have a sum of the middle term coefficient, 3. The chart below shows all the possible factors of −18 and the sum of each pair of factors.

Factors of −18	*Sum of Factors*
1, −18	−17
−1, 18	17
2, −9	−7
−2, 9	7
3, −6	−3
−3, 6	3

Step 3: Rewrite the middle term of the trinomial using the pair of factors whose sum is equal to 3. The order of the two terms does not matter.

$$2x^2 - 3x + 6x - 9$$

Step 4: Group the first two terms together and the last two terms together using parentheses. Notice how the sign of the third term is outside of the parentheses.

$$(2x^2 - 3x) + (6x - 9)$$

Step 5: Factor out the greatest common factor of each pair of parentheses.

$$x(2x - 3) + 3(2x - 3)$$

Step 6: Notice that the binomial expressions in parentheses are the same. Factor out the common binomial to obtain the answer.

$$(2x - 3)(x + 3)$$

Example 26: Factor $4x^2 + 16x + 7$.

First, the product of $4(7) = 28$. Then, two factors of 28 whose sum is 16 are 2 and 14. Rewrite the middle term of the trinomial using $2x$ and $14x$.

$$4x^2 + 2x + 14x + 7$$

Then, group the pairs using parentheses and factor out a GCF in each pair.

$$(4x^2 + 2x) + (14x + 7)$$

$$2x(2x + 1) + 7(2x + 1)$$

Lastly, factor out the common binomial to obtain the solution.

$$(2x + 1)(2x + 7)$$

Summary of the factoring methods

When factoring polynomials, you should look for factoring in the following order.

1. Look for the greatest common factor if one exists.
2. If there are two terms, look for the difference of square numbers.
3. If there are three terms, look for a pattern that applies to $ax^2 + bx + c$ or factor by grouping.
4. If there are four or more terms, look for some type of regrouping that will lead to other factoring.

There are polynomials that are *not* factorable, such as the one in Example 27.

Example 27: Factor $2x^2 + 3x + 5$.

1. This polynomial does not have a common factor.
2. This polynomial is not a difference of square numbers.
3. There is no $(\_x \quad)(\_x \quad)$ combination that produces $2x^2 + 3x + 5$.
4. Since there are only three terms, there is no regrouping possibility.

Therefore, this polynomial is not factorable.

Chapter Check-Out

Questions

1. Simplify $mr^2 + 3mr^2 - 5mr^2$. Find the degree of the simplified monomial.
2. Simplify $(4xy^4)^3(x^2y^3)$.
3. Show that $\left(4\frac{1}{2}\right)^2 \neq 16\frac{1}{4}$ using the distributive property.
4. Simplify $\frac{2x^2y^4z^2}{10xy^3z}$.
5. Write $\frac{a}{b^{-3}}$ in simplest form.
6. Simplify $(5a^{-3}b)(7a^{-2}b)$.
7. Explain why the algebraic expression $4x^2 - 5\sqrt{x} + \sqrt{3}$ is not a polynomial.
8. What is the result when $(3x^2 - 6x + 8)$ is subtracted from $(7x^2 + 2x - 4)$? Express the result as a trinomial in standard form.
9. The length of a rectangle is represented by $3x - 2$ and the width by $2x + 1$. Express the area of the rectangle as a single polynomial in simplest form if $A = lw$.
10. Find the value of b so that the polynomial $(2x^2 - bx - 3)$ is divisible by $(x - 3)$.
11. Factor $25x^2 - 100$ completely.
12. Find two different coefficients for the x term below so that the trinomial can be factored into a product of two binomials. Then, factor each trinomial.

$$x^2 + \_x - 15$$

13. You are factoring the trinomial $2x^2 - bx - 15$ using grouping, then rewriting the polynomial as $2x^2 - px + qx - 15$. If $b = 7$, find p and q, then continue the process of factoring by grouping. What is the value of pq?
14. Factor: $m^2 - 2mn - 3n^2$.
15. Factor: $xy + 3y + x + 3$.

16. Name the polynomial below based on its degree and number of terms. Factor the polynomial completely.

$$8x^2 + 26x + 20$$

For question 17, select the best answer from among the four choices.

17. The quotient of two polynomials will not always result in another polynomial. This means that polynomials are not closed under division. Which quotient below verifies this statement? Explain your answer.

A. $\dfrac{9x^3+4x}{2x}$

B. $(4x^2-8x+3)\div(2x-1)$

C. $\dfrac{12x^8+x^5+3x}{3x^3}$

D. $\dfrac{2x^2+4x}{x+2}$

18. Explain why it is not possible to factor the polynomial $x^2 - 3x + 12$.

19. Find the error in factoring the following trinomial.

$$x^2 + 5x - 6 = (x + 2)(x + 3)$$

Answers

1. $-mr^2$; the degree of the monomial is 3.

2. $64x^5y^{15}$

3. $$\left(4\frac{1}{2}\right)^2=\left(4+\frac{1}{2}\right)^2$$
$$=\left(4+\frac{1}{2}\right)\left(4+\frac{1}{2}\right)$$
$$=4\left(4+\frac{1}{2}\right)+\frac{1}{2}\left(4+\frac{1}{2}\right)$$
$$=16+2+2+\frac{1}{4}$$
$$=20\frac{1}{4}$$

4. $\frac{xyz}{5}$
5. ab^3
6. $\frac{35b^2}{a^5}$
7. For an algebraic expression to be a polynomial, all the variables in the expression must have whole-number exponents. The term $5\sqrt{x}$ is equivalent to $5x^{\frac{1}{2}}$, which is not a whole-number exponent. Therefore, the algebraic expression $4x^2 - 5\sqrt{x} + \sqrt{3}$ is not a polynomial.
8. $4x^2 + 8x - 12$
9. $6x^2 - x - 2$
10. $b = 5$
11. $25(x - 2)(x + 2)$
12. The coefficient of the x term can either be 14 or 2. $x^2 + 14x - 15$ factors into $(x - 1)(x + 15)$ and $x^2 + 2x - 15$ factors into $(x - 3)(x + 5)$.
13. $pq = -30$. So, $p = 10$ and $q = 3$. Continuing the process of factoring by grouping gives $2x^2 - 10x + 3x - 15 = (2x^2 - 10x) + (3x - 15) = 2x(x - 5) + 3(x - 5) = (2x + 3)(x - 5)$.
14. $(m - 3n)(m + n)$
15. $(y + 1)(x + 3)$
16. Second-degree trinomial; $2(4x + 5)(x + 2)$
17. The correct answer is C. When the polynomial $12x^8 + x^5 + 3x$ is divided by the monomial $3x^3$, the quotient is $4x^5 + \frac{1}{3}x^2 + \frac{1}{x^2}$. Since $\frac{1}{x^2} = x^{-2}$, the algebraic expression is not a polynomial, verifying that polynomials are not closed under division.
18. The polynomial $x^2 - 3x + 12$ can't be factored because there are no factors of 12 whose sum is -3.
19. To factor the trinomial, $x^2 + 5x - 6$, you need to find factors of 6 whose difference is 5. While 2 and 3 are factors of 6, their difference is not 5. The correct factors are 1 and 6, since their difference is 5. Therefore, the correct factoring is $(x + 6)(x - 1)$.

Chapter 6

RATIONAL EXPRESSIONS

Chapter Check-In

- ❑ Simplifying rational expressions
- ❑ Multiplying rational expressions
- ❑ Dividing rational expressions
- ❑ Adding or subtracting rational expressions

Rational expressions are fractions in which the numerator and/or the denominator are polynomials, such as $\frac{3}{x}$. Because division by 0 is undefined, values of the variable in the denominator have certain restrictions. Since the denominator can *never* equal 0, these values are called **excluded values.** Therefore, in the fraction

$\frac{5}{x}$	x cannot equal 0 $(x \neq 0)$
$\frac{2}{x-3}$	x cannot equal 3 $(x \neq 3)$
$\frac{3}{a-b}$	$a-b$ cannot equal 0 $(a-b \neq 0)$ so a cannot equal b $(a \neq b)$
$\frac{4}{a^2b}$	neither a nor b can equal 0 $(a \neq 0, b \neq 0)$

Be aware of these types of restrictions. Note that you find excluded values in the original expression before you simplify a rational expression.

Operations with Rational Expressions

Common Core Standard: Arithmetic with Polynomials and Rational Expressions

Understand that rational expressions form a system analogous to the rational numbers, closed under addition, subtraction, multiplication, and division by a nonzero rational expression; add, subtract, multiply, and divide rational expressions (A.APR.7).

Many techniques will simplify your work as you perform operations with rational expressions. As you review the examples in this chapter, note the steps involved in each operation and any methods that will save you time. Keep in mind that rational expressions can be written in many different ways.

Simplifying rational expressions

To simplify a rational expression, first factor the numerator and the denominator; then **cancel** (or divide out) common factors. Remember, just like the simplified form of a numerical fraction, the numerator and denominator of a rational expression should have no common factors other than 1.

Example 1: Simplify and state any excluded values.

(a) $\dfrac{4x^3}{8x^2} = \dfrac{\overset{1}{\cancel{4}}}{\underset{2}{\cancel{8}}} \dfrac{\overset{x}{\cancel{x^3}}}{\underset{1}{\cancel{x^2}}} = \dfrac{1}{2}x \text{ or } \dfrac{x}{2},\ x \neq 0$

(b) $\dfrac{3x-3}{4x-4} = \dfrac{3(x-1)}{4(x-1)} = \dfrac{3\overset{1}{\cancel{(x-1)}}}{4\underset{1}{\cancel{(x-1)}}} = \dfrac{3}{4},\ x \neq 1$

(c) $\dfrac{x^2+2x+1}{3x+3} = \dfrac{(x+1)(x+1)}{3(x+1)} = \dfrac{\overset{1}{\cancel{(x+1)}}(x+1)}{3\underset{1}{\cancel{(x+1)}}} = \dfrac{x+1}{3},\ x \neq -1$

Warning: Do NOT cancel terms through an addition or subtraction sign as shown on the next page.

$$\frac{x+1}{x+2} \neq \frac{\cancel{x}+1}{\cancel{x}+2} \neq \frac{1}{2}$$

or

$$\frac{x+6}{6} \neq \frac{x+\cancel{6}}{\cancel{6}} \neq x$$

Multiplying rational expressions

To multiply rational expressions, first factor the numerators and denominators that are polynomials. Then, just as with multiplying numerical fractions, simplify where possible. Multiply the remaining numerators together and denominators together. (If you've simplified properly, your answer will be in simplest form.)

Example 2: Multiply.

(a) $\frac{2x}{3} \cdot \frac{y}{5} = \frac{2x \cdot y}{3 \cdot 5} = \frac{2xy}{15}$

(b) $\frac{x^2}{3y} \cdot \frac{2y}{3x} = \frac{\overset{1}{x^{\cancel{2}}}}{3\underset{1}{\cancel{y}}} \cdot \frac{2\overset{1}{\cancel{y}}}{3\underset{1}{\cancel{x}}} = \frac{2x}{9}$

(c) $\frac{x+1}{5y+10} \cdot \frac{y+2}{x^2+2x+1} = \frac{x+1}{5(y+2)} \cdot \frac{y+2}{(x+1)(x+1)}$

$$= \frac{\overset{1}{\cancel{(x+1)}}}{5\underset{1}{\cancel{(y+2)}}} \cdot \frac{\overset{1}{\cancel{(y+2)}}}{\underset{1}{\cancel{(x+1)}}(x+1)}$$

$$= \frac{1}{5(x+1)}$$

(d) $\frac{x^2-4}{6} \cdot \frac{3y}{2x+4} = \frac{(x+2)(x-2)}{6} \cdot \frac{3y}{2(x+2)}$

$$= \frac{\overset{1}{\cancel{(x+2)}}(x-2)}{\underset{2}{\cancel{6}}} \cdot \frac{\overset{1}{\cancel{3}}y}{2\underset{1}{\cancel{(x+2)}}}$$

$$= \frac{(x-2)y}{4}$$

(e) $\frac{x^2+4x+4}{x-3}\cdot\frac{5}{3x+6}=\frac{(x+2)(x+2)}{(x-3)}\cdot\frac{5}{3(x+2)}$

$$=\frac{(x+2)\cancel{(x+2)}^{1}}{(x-3)}\cdot\frac{5}{3\cancel{(x+2)}_{1}}$$

$$=\frac{5(x+2)}{3(x-3)}$$

Dividing rational expressions

To divide rational expressions, multiply the first fraction (the dividend) by the reciprocal of the second fraction (the divisor). Remember, you can simplify only after you take the reciprocal.

Example 3: Divide.

(a) $\frac{3x^2}{5}\div\frac{2x}{y}=\frac{3x^2}{5}\cdot\frac{y}{2x}=\frac{3x^{\cancel{2}\,1}}{5}\cdot\frac{y}{2\cancel{x}_{1}}=\frac{3xy}{10}$

(b) $\frac{4x-8}{6}\div\frac{x-2}{3}=\frac{4x-8}{6}\cdot\frac{3}{x-2}$

$$=\frac{4(x-2)}{6}\cdot\frac{3}{x-2}$$

$$=\frac{4\cancel{(x-2)}^{1}}{\cancel{6}_{2}}\cdot\frac{\cancel{3}^{1}}{\cancel{x-2}_{1}}$$

$$=\frac{4}{2}=2$$

Adding or subtracting rational expressions

To add or subtract rational expressions that have a common denominator, simply keep the denominator and combine (add or subtract) the numerators. Simplify if possible.

Example 4: Perform the indicated operation.

(a) $\frac{4}{x}+\frac{5}{x}=\frac{4+5}{x}=\frac{9}{x}$

(b) $\frac{x-4}{x+1}+\frac{3}{x+1}=\frac{x-4+3}{x+1}=\frac{x-1}{x+1}$

(c) $\frac{3x}{y}-\frac{2x-1}{y}=\frac{3x-(2x-1)}{y}=\frac{3x-2x+1}{y}=\frac{x+1}{y}$

To add or subtract algebraic fractions that have different denominators, first find a least common denominator (LCD), change each fraction to an equivalent fraction with the common denominator, and then combine each numerator. Simplify if possible.

Example 5: Perform the indicated operation.

(a) $\frac{2}{x}+\frac{3}{y}=$

$\text{LCD}=xy$

$\frac{2}{x}\cdot\frac{y}{y}+\frac{3}{y}\cdot\frac{x}{x}=\frac{2y}{xy}+\frac{3x}{xy}=\frac{2y+3x}{xy}$

(b) $\frac{x+2}{3x}+\frac{x-3}{6x}=$

$\text{LCD}=6x$

$\frac{x+2}{3x}\cdot\frac{2}{2}+\frac{x-3}{6x}=\frac{2x+4}{6x}+\frac{x-3}{6x}=\frac{2x+4+x-3}{6x}=\frac{3x+1}{6x}$

If there is a common variable factor with more than one exponent, use the largest exponent of that variable.

Example 6: Perform the indicated operation.

(a) $\frac{2}{y^2}-\frac{3}{y}=$

$\text{LCD}=y^2$

$\frac{2}{y^2}-\frac{3}{y}\cdot\frac{y}{y}=\frac{2}{y^2}-\frac{3y}{y^2}=\frac{2-3y}{y^2}$

(b) $\dfrac{4}{x^3y}+\dfrac{3}{xy^2}=$

$\text{LCD}=x^3y^2$

$$\frac{4}{x^3y}\cdot\frac{y}{y}+\frac{3}{xy^2}\cdot\frac{x^2}{x^2}=\frac{4y}{x^3y^2}+\frac{3x^2}{x^3y^2}=\frac{4y+3x^2}{x^3y^2}$$

(c) $\dfrac{x}{x+1}-\dfrac{2x}{x+2}=$

$\text{LCD}=(x+1)(x+2)$

$$\frac{x}{x+1}\cdot\frac{x+2}{x+2}-\frac{2x}{x+2}\cdot\frac{x+1}{x+1}=\frac{x^2+2x}{(x+1)(x+2)}-\frac{2x^2+2x}{(x+1)(x+2)}$$

$$=\frac{x^2+2x-2x^2-2x}{(x+1)(x+2)}$$

$$=\frac{-x^2}{(x+1)(x+2)}$$

Note: In Example 6, item (c), distribute the negative sign to the entire numerator in the second fraction.

To find the LCD, it is often necessary to factor the denominators and proceed as follows.

Example 7: Perform the indicated operation.

$$\frac{2x}{x^2-9}-\frac{5}{x^2+4x+3}=\frac{2x}{(x+3)(x-3)}-\frac{5}{(x+3)(x+1)}$$

$$\text{LCD}=(x+3)(x-3)(x+1)$$

$$\frac{2x}{(x+3)(x-3)}\cdot\frac{(x+1)}{(x+1)}-\frac{5}{(x+3)(x+1)}\cdot\frac{(x-3)}{(x-3)}=$$

$$\frac{2x^2+2x}{(x+3)(x-3)(x+1)}-\frac{5x-15}{(x+3)(x-3)(x+1)}=$$

$$\frac{2x^2+2x-(5x-15)}{(x+3)(x-3)(x+1)}=$$

$$\frac{2x^2+2x-5x+15}{(x+3)(x-3)(x+1)}=\frac{2x^2-3x+15}{(x+3)(x-3)(x+1)}$$

Occasionally, a problem will require simplifying what appears to be the final result, as in Example 8.

Example 8: Perform the indicated operation.

$$\frac{x}{x^2+5x+6}-\frac{2}{x^2+3x+2}=\frac{x}{(x+3)(x+2)}-\frac{2}{(x+2)(x+1)}$$

$$\text{LCD}=(x+1)(x+2)(x+3)$$

$$\frac{x}{(x+3)(x+2)}\cdot\frac{(x+1)}{(x+1)}-\frac{2}{(x+2)(x+1)}\cdot\frac{(x+3)}{(x+3)}=$$

$$\frac{x^2+x}{(x+1)(x+2)(x+3)}-\frac{2x+6}{(x+1)(x+2)(x+3)}=$$

$$\frac{x^2+x-2x-6}{(x+1)(x+2)(x+3)}=$$

$$\frac{x^2-x-6}{(x+1)(x+2)(x+3)}=$$

$$\frac{(x-3)(x+2)}{(x+1)(x+2)(x+3)}=$$

$$\frac{(x-3)\,\overset{1}{\cancel{(x+2)}}}{(x+1)\,\underset{1}{\cancel{(x+2)}}\,(x+3)}=\frac{(x-3)}{(x+1)(x+3)}$$

Chapter Check-Out

Questions

1. Simplify $\frac{9x^2}{12x^3}$.

2. The rational expression $\frac{t}{x^2-x-12}$ simplifies to $\frac{x-5}{x+3}$. Write the trinomial that represents t.

3. Find the product of $\frac{x-1}{x}\cdot\frac{x^2+3x}{x^2-7x+6}$. State any excluded values.

4. Find the values of x in which the expression $\frac{2x}{x-2} \div \frac{2x^3-50x}{x^2+3x-10}$ is undefined. Explain your answer.

5. Find the quotient of $\frac{10y+5}{4} \div \frac{2y+1}{2}$.

6. Explain the error that was made in the equation below when adding the rational expressions.

$$\frac{7}{x}+\frac{3}{y}=\frac{10}{x+y}$$

7. Find the sum of $\frac{3}{x^4y^2}+\frac{2}{x^2y^3}$.

8. Find the difference of $\frac{2x}{x-1}-\frac{x}{x+2}$.

9. Find the difference of $\frac{x}{x^2+11x+30}-\frac{5}{x^2+9x+20}$.

Question 10 refers to the following figure.

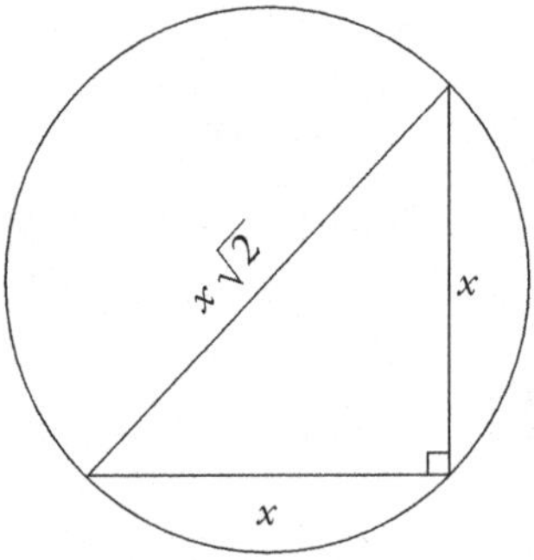

10. Write the ratio representing the area of the triangle to the area of the circle above. Write your answer in terms of π.

Questions 11 and 12 refer to the following information.

A cross-county skier travels 3 miles up a slight hill to his cabin. The next day, he travels back down the slope 20 percent faster. Let r represent the rate the skier travels uphill.

11. Write an expression that represents the time the skier travels both ways.

12. If it took the skier 1.5 hours to travel both ways, at what rate did the skier travel downhill, and approximately how long did it take him to travel down the hill?

Answers

1. $\frac{3}{4x}$

2. $x^2 - 9x + 20$

3. $\frac{x+3}{x-6}$; $x \neq 6, x \neq 0, x \neq 1$

4. 2, −5, 5, and 0 are excluded values because they make the denominators equal to zero.

5. $\frac{5}{2}$

6. The denominators were combined instead of finding the least common denominator to add the rational expressions.

7. $\frac{3y+2x^2}{x^4y^3}$

8. $\frac{x^2+5x}{x^2+x-2}$ or $\frac{x(x+5)}{x^2+x-2}$

9. $\frac{x-6}{(x+4)(x+6)}$

10. $\frac{1}{\pi}$

11. Using the distance formula, $d = rt$, solve for t: $t = \frac{d}{r}$. Since the skier travels downhill 20 percent faster, $1.2r$ represents the rate the skier travels downhill. Therefore, the total time $= \frac{3}{r} + \frac{3}{1.2r} = \frac{3}{r} \cdot \frac{1.2}{1.2} + \frac{3}{1.2r} = \frac{3.6}{1.2r} + \frac{3}{1.2r} = \frac{6.6}{1.2r} = \frac{5.5}{r}$.

12. If the total time it took the skier to travel uphill and downhill was 1.5 hours, substitute 1.5 for total time and solve for *r*.

$$\text{total time} = \frac{3}{r} + \frac{3}{1.2r} = \frac{5.5}{r}$$

To find the rate the skier traveled downhill,

$$1.5 = \frac{5.5}{r}$$

$$r = 3.\bar{6} \text{ mph}$$

The skier traveled uphill at a rate of $3.\bar{6}$ mph. Therefore, the skier traveled downhill $1.2r = 1.2(3.66666666) = 4.4$ mph. It took the skier about 41 minutes to travel downhill since $\frac{3}{1.2r} = \frac{3}{1.2(3.\bar{6})} = 0.6\overline{81}$ hrs $\times$ 60 min $\approx$ 41 min.

Chapter 7

INEQUALITIES, GRAPHING, AND ABSOLUTE VALUE

Chapter Check-In

- ❑ Inequalities and their properties
- ❑ Solving inequalities
- ❑ Graphing inequalities on a number line
- ❑ Intervals and Interval Notation
- ❑ Absolute value
- ❑ Solving equations containing an absolute value
- ❑ Solving inequalities containing an absolute value

Now that you know how to analyze equations and monomials and can perform the sequence of steps to arrive at a solution, you can extend your earlier work to create and graph inequalities. Common Core Mathematics focuses on applying reasoning skills to interpret and explain viable options to arrive at inequality solutions.

Inequalities

Common Core Standard: Creating Equations

Create equations and inequalities in one variable and use them to solve problems (A.CED.1). Represent constraints by equations or inequalities, and by systems of equations and/or inequalities, and interpret solutions as viable or non-viable options in a modeling context (A.CED.3).

Common Core Standard: Reasoning with Equations and Inequalities

Solve linear equations and inequalities in one variable, including equations with coefficients represented by letters (A.REI.3).

An **inequality** is a statement in which two mathematical expressions are not equal. Instead of using an equal sign (=) as in an equation, inequality symbols are used:

$>$ (is greater than)

$<$ (is less than)

$\geq$ (is greater than or equal to)

$\leq$ (is less than or equal to)

Axioms and properties of inequality

Just as the properties of equality are used to solve equations, you can apply the properties of inequality to justify the steps being taken to solve an inequality. For all real numbers a, b, and c, the following are some basic axioms and properties of inequality:

- **Trichotomy axiom:** $a > b$, $a = b$, or $a < b$.

  These are the only possible relationships between two numbers. Either the first number is greater than the second, the numbers are equal, or the first number is less than the second.

- **Transitive axiom:** If $a > b$ and $b > c$, then $a > c$.

  Therefore, if $3 > 2$ and $2 > 1$, then $3 > 1$.

  If $a < b$ and $b < c$, then $a < c$.

  Therefore, if $4 < 5$ and $5 < 6$, then $4 < 6$.

- **Addition and subtraction property:**

  If $a > b$, then $a + c > b + c$.

  If $a > b$, then $a - c > b - c$.

  If $a < b$, then $a + c < b + c$.

  If $a < b$, then $a - c < b - c$.

  Adding or subtracting the same amount from each side of an inequality keeps the direction of the inequality the same.

  For example:
  If $3 > 2$, then $3 + 1 > 2 + 1$ (that is, $4 > 3$).

  If $12 < 15$, then $12 - 4 < 15 - 4$ (that is, $8 < 11$).

- **Positive multiplication and division property:**

  If $a > b$ and $c > 0$, then $ac > bc$.

  If $a < b$ and $c > 0$, then $ac < bc$.

  If $a > b$ and $c > 0$, then $\frac{a}{c} > \frac{b}{c}$.

  If $a < b$ and $c > 0$, then $\frac{a}{c} < \frac{b}{c}$.

  Multiplying or dividing each side of an inequality by a positive number keeps the direction of the inequality the same.

  For example:

  If $5 > 2$, then $5(3) > 2(3)$; therefore, $15 > 6$.

  If $3 < 12$, then $\frac{3}{4} < \frac{12}{4}$, or $\frac{3}{4} < 3$.

- **Negative multiplication and division property:**

  If $a > b$ and $c < 0$, then $ac < bc$.

  If $a < b$ and $c < 0$, then $ac > bc$.

  If $a > b$ and $c < 0$, then $\frac{a}{c} < \frac{b}{c}$.

  If $a < b$ and $c < 0$, then $\frac{a}{c} > \frac{b}{c}$.

  Multiplying or dividing each side of an inequality by a negative number reverses the direction of the inequality.

  For example:

  If $5 > 2$, then $5(-3) < 2(-3)$; therefore, $-15 < -6$.

  If $3 < 12$, then $\frac{3}{-4} > \frac{12}{-4}$, or $-\frac{3}{4} > -3$.

Solving inequalities

When solving inequalities in one variable, treat them exactly like equations, with one exception: If you multiply or divide each side of the inequality by a negative number, you must reverse the direction of the inequality symbol. The solution of an inequality is any value substituted for the variable that will make the inequality *true*. This is referred to as the solution set, and can be written as an inequality, in set-builder notation, in interval notation, or as a graph on a number line.

Example 1: Solve for x: $2x + 4 > 6$.

$$\begin{array}{rl} 2x+4 > & 6 \\ -4 & -4 \\ \hline 2x \quad > & 2 \end{array}$$

$$\frac{2x}{2} > \frac{2}{2}$$

$$x > 1$$

Note that the solution to this inequality has infinitely many solutions. To check the solution $x > 1$, any number greater than 1 can be used.

$$\begin{array}{lrcr} \text{Check:} & x = 2 & \text{or} & x = 8 \\ & 2x+4>6 & & 2x+4>6 \\ & 2(2)+4>6 & & 2(8)+4>6 \\ & 8>6\checkmark & & 20>6\checkmark \end{array}$$

The solution of an inequality can sometimes be written in **set-builder notation** $\{x|x > 1\}$, which is read "the set of all values of x such that x is greater than 1."

Example 2: Solve for x: $-7x > 14$.

Divide by -7 and reverse the direction of the inequality symbol.

$$\frac{-7x}{-7} < \frac{14}{-7}$$

$$x < -2$$

Example 3: Solve for x: $3x + 2 \geq 5x - 10$.

$$\begin{array}{rl} 3x+2 \geq & 5x-10 \\ -2 & \quad -2 \\ \hline 3x \quad \geq & 5x-12 \end{array}$$

$$\begin{array}{rcr} 3x & \geq & 5x-12 \\ -5x & & -5x \\ \hline -2x & \geq & -12 \end{array}$$

$$\frac{-2x}{-2} \leq \frac{-12}{-2}$$

$$x \leq 6$$

Notice that inverse operations are used in Example 3. Divide each side of the inequality by -2 and reverse the direction of the inequality symbol.

In set-builder notation, $\{x|x \leq 6\}$.

Graphing on a Number Line

Integers and real numbers can be represented on a number line. The point on this line associated with each number is called the *graph* of the number. Take a look at the number lines shown below. Notice that the numbers are spaced equally, or proportionately.

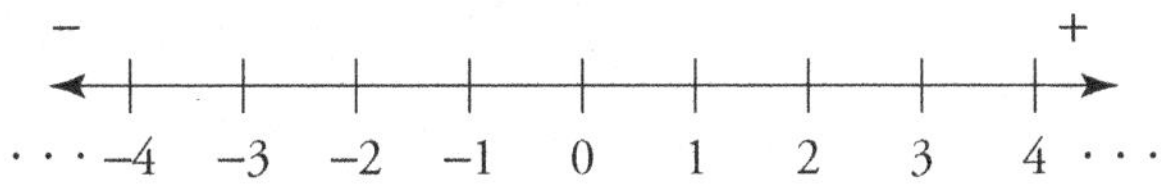

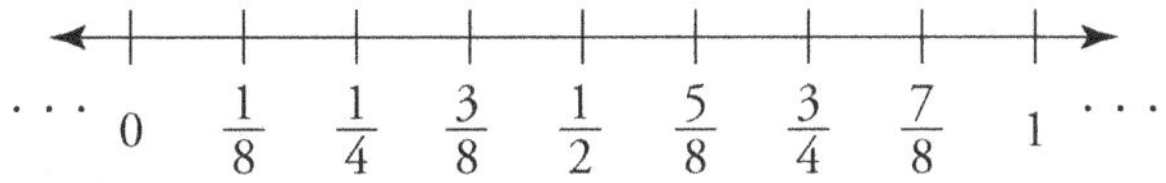

Graphing inequalities

Information and solutions about inequalities are found by analyzing graphs. When graphing inequalities involving only integers, dots are used. The solution of an inequality can be graphed on a number line.

Example 4: Graph the set of all values of x such that $1 \leq x \leq 4$ and x is an integer.

$$\{x|1 \leq x \leq 4, x \text{ is an integer}\}$$

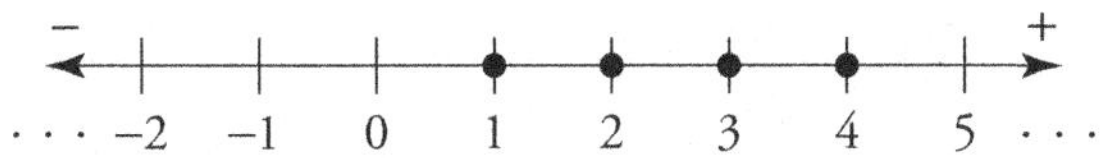

When graphing inequalities involving real numbers, lines, rays, and dots are used. A closed dot is used if the number is included. A hollow dot is used if the number is not included.

Example 5: Graph as indicated.

(a) Graph the set of all values of x such that $x \geq 1$.

$$\{x|x \geq 1\}$$

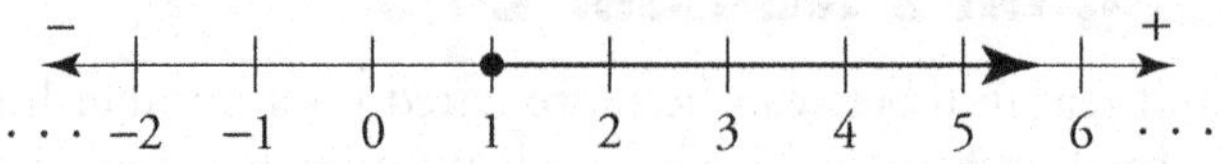

This ray is often called a **closed ray** or a *closed half-line.*

(b) Graph the set of all values of x such that $x > 1$.

$$\{x|x > 1\}$$

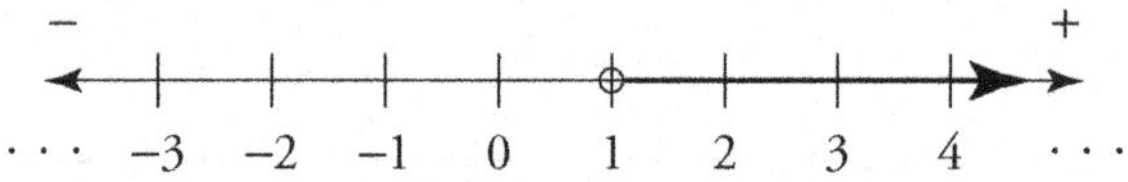

(c) Graph the set of all values of x such that $x < 4$.

$$\{x|x < 4\}$$

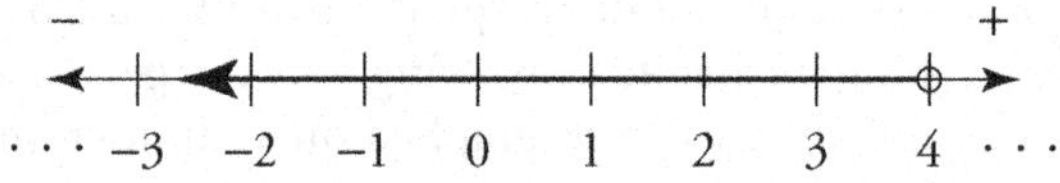

The rays in Examples 5(b) and 5(c) are often called **open rays** or *open half-lines.* The hollow dot distinguishes an open ray from a closed ray.

Interval

An **interval** consists of all the values that lie within two certain boundaries. You can write the solution of an inequality in **interval notation** to describe the range of values that would make an inequality true. This notation uses an interval to represent a pair of boundary numbers with brackets and/or parentheses. Brackets are used to show the value is part of the solution, whereas parentheses are used to show the value is not part of the solution. If the two boundaries, or endpoints, are included, then the interval is called a **closed interval.** For example, the inequality $0 \le x \le 5$ is written as [0, 5]. If the endpoints are not included, then the interval is called an **open interval.** For example, the inequality $0 < x < 5$ is written as (0, 5).

Example 6:

(a) Graph the closed interval [−1, 2].

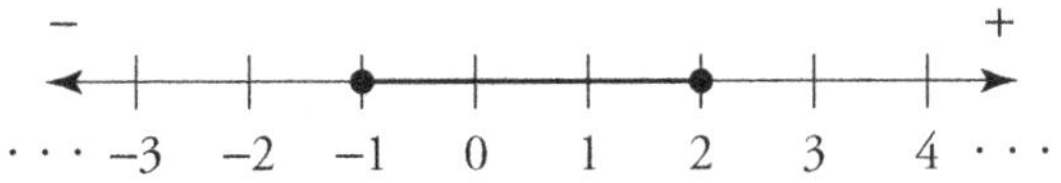

(b) Graph the open interval (−2, 2).

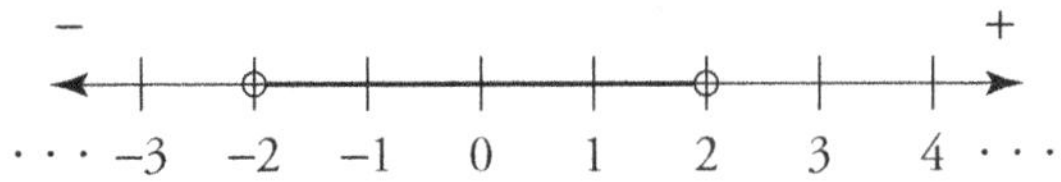

If one endpoint is included and one is not, then the interval is called a **half-open interval.** For example, the inequality $0 \le x < 5$ is written as [0, 5).

Example 7: Graph the half-open interval (−1, 2].

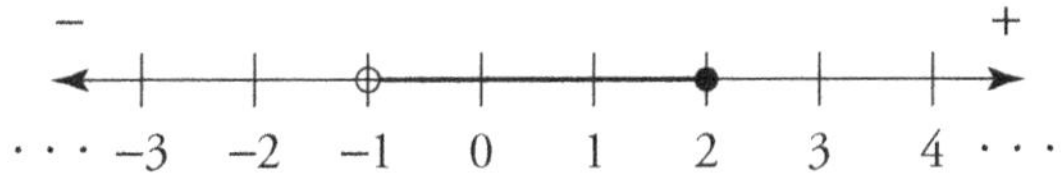

If the values to one side of an endpoint extend on forever, then the interval is called a **non-ending interval.** For example, the inequality $x \le 4$ is written as $(-\infty, 4]$.

Example 8: Graph the non-ending interval $(-\infty, -4]$

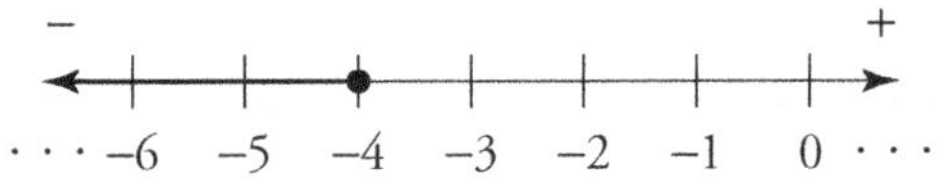

The solution of an inequality in one variable can be written as an inequality, in set-builder notation, in interval notation, or graphed on a number line, as shown in the table below.

Inequality	*Set-Builder Notation*	*Interval Notation*	*Graph*
$x > -3$	$\{x \mid x > -3\}$	$(-3, \infty)$	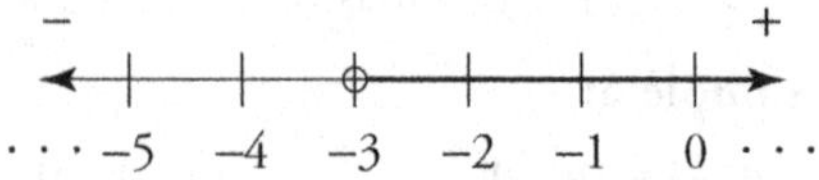

Absolute Value

The numerical value when direction or sign is not considered is called the **absolute value.** The absolute value of x is written $|x|$. Geometrically, the absolute value of a number is its distance from zero on a number line. Since distance on a number line is nonnegative, the absolute value of a number is always nonnegative except when the number is 0.

$$|0| = 0$$

$$|x| > 0 \text{ when } x \neq 0$$

$$|-x| > 0 \text{ when } x \neq 0$$

Example 9: Simplify.

(a) $|4| = 4$

(b) $|-6| = 6$

(c) $|7 - 9| = |-2| = 2$

(d) $3 - |-6| = 3 - 6 = -3$

Note that work inside absolute value bars should be done before performing any other operations.

Solving equations containing an absolute value

To solve an equation containing an absolute value, isolate the absolute value expression on one side of the equation. Then set the expression inside the absolute value bars equal to both the positive and negative values of the number on the other side of the equation and solve both equations. Examples 10 and 11 provide a detailed explanation based on the definition of absolute value.

Example 10: Solve $|x| + 2 = 5$.

Isolate the absolute value.

$$\begin{array}{rcr} |x| + 2 &=& 5 \\ -2 && -2 \\ \hline |x| &=& 3 \end{array}$$

Since there are two values that are 3 units from zero on the number line, the equation has two solutions: 3 and -3. Therefore, $x = 3$ or $x = -3$.

Example 11: Solve $3|x - 1| - 1 = 11$. Graph the solutions.

Isolate the absolute value expression.

$$\begin{array}{rcr} 3|x-1| - 1 &=& 11 \\ +1 && +1 \\ \hline 3|x-1| &=& 12 \end{array}$$

$$\frac{3|x-1|}{3} = \frac{12}{3}$$

$$|x-1| = 4$$

Since the absolute value equation contains a variable expression within the absolute value bars, the equation $|x - 1| = 4$ means that the distance from $x - 1$ to zero on a number line is 4 units. Since there are two values that are 4 units from zero on a number line, set the expression inside the absolute value bars equal to 4 and -4.

Solving for x,

$$\begin{array}{rcr} x - 1 &=& 4 \\ +1 && +1 \\ \hline x &=& 5 \end{array} \quad \text{or} \quad \begin{array}{rcr} x - 1 &=& -4 \\ +1 && +1 \\ \hline x &=& -3 \end{array}$$

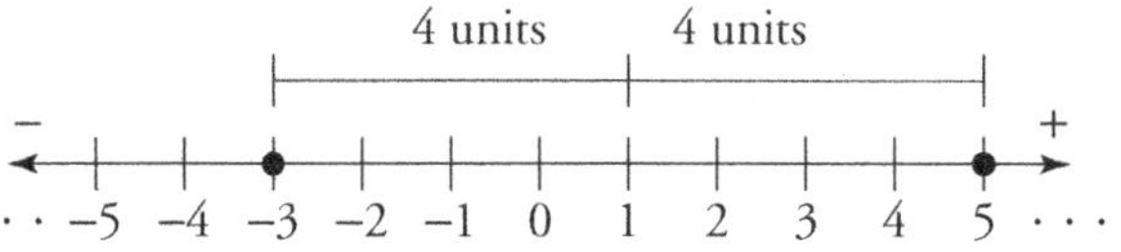

Note that each solution is 4 units from 1.

Solving inequalities containing an absolute value and graphing

To solve an inequality containing an absolute value, begin with the same steps as for solving equations with an absolute value. When writing two separate inequalities by comparing the expression inside the absolute value bars to both the + and − of the other side of the inequality, reverse the direction of the inequality symbol when writing the inequality with the negative. Note that the solution to an absolute value inequality may be a **compound inequality,** which is when two inequalities are joined by the word *and* or the word *or.*

Example 12: Solve and graph the solution: $|x - 1| > 2$.

Notice that the absolute value expression is already isolated.

$$|x - 1| > 2$$

Write two separate inequalities by comparing the expression inside the absolute value bars to both 2 and −2. Be sure to reverse the direction of the inequality when comparing it with −2.

Solve for x.

$$\begin{array}{rcr} x - 1 > 2 & \text{or} & x - 1 < -2 \\ +1 \quad +1 & & +1 \quad +1 \\ \hline x \quad > 3 & \text{or} & x \quad < -1 \end{array}$$

Graph the answer.

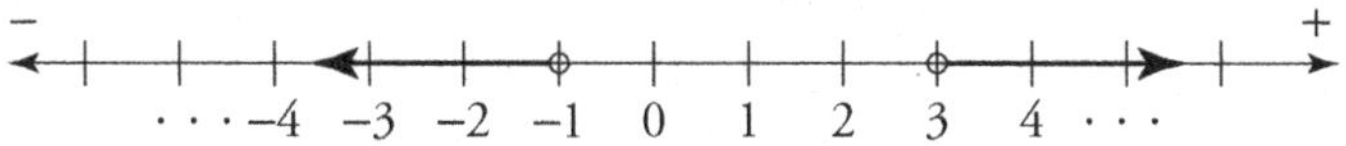

Example 13: Solve and graph the solution: $3|x| - 2 \leq 1$.

Isolate the absolute value.

$$\begin{array}{r} 3|x| - 2 \leq 1 \\ +2 \quad +2 \\ \hline 3|x| \quad \leq 3 \end{array}$$

$$\frac{3|x|}{3} \leq \frac{3}{3}$$

$$|x| \leq 1$$

Write two separate inequalities by comparing the expression inside the absolute value bars to both 1 and -1. Be sure to reverse the direction of the inequality symbol when comparing it with -1.

$$x \leq 1 \text{ and } x \geq -1$$

This inequality can also be written as $-1 \leq x \leq 1$.

Graph the solution.

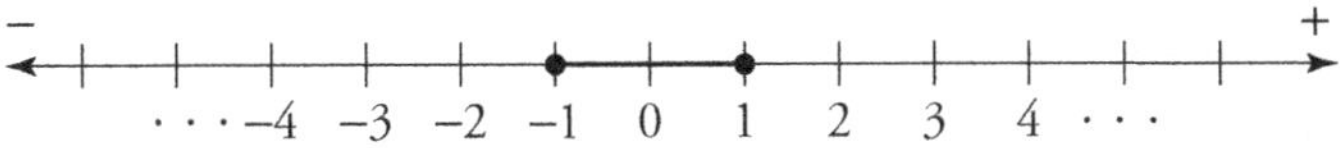

Example 14: Solve and graph the solution: $2|1 - x| + 1 \geq 3$.

Isolate the absolute value.

$$\begin{array}{rcl} 2|1-x|+1 & \geq & 3 \\ -1 & & -1 \\ \hline 2|1-x| & \geq & 2 \end{array}$$

$$\frac{2|1-x|}{2} \geq \frac{2}{2}$$

$$|1-x| \geq 1$$

Write two separate inequalities by comparing the expression inside the absolute value symbol to both 1 and -1. Be sure to reverse the direction of the inequality symbol when comparing it with -1.

Solve for x. (Remember to switch the direction of the inequality symbol when dividing by a negative.)

$$\begin{array}{rcl} 1-x \geq 1 & \text{or} & 1-x \leq -1 \\ -1 \quad\; -1 & & -1 \quad\; -1 \\ \hline -x \;\; \geq 0 & & -x \leq -2 \end{array}$$

$$\frac{-x}{-1} \leq \frac{0}{-1} \qquad \frac{-x}{-1} \geq \frac{-2}{-1}$$

$$x \leq 0 \text{ or } x \geq 2$$

Graph the solution.

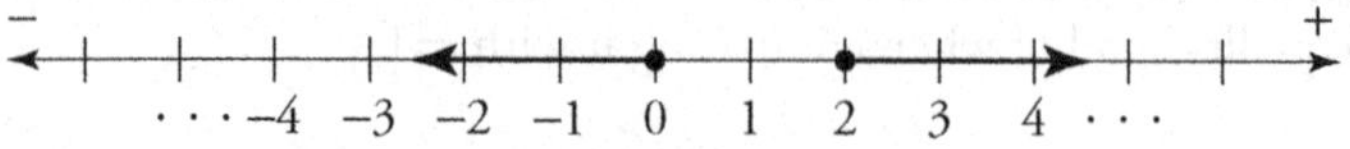

Remember, using the word *and* denotes the intersection of two sets, and using the word *or* denotes the union of two sets (see Chapter 2, "Structure in Algebraic Expressions").

Chapter Check-Out

Questions

1. True or false: If $a > b$ and $b > c$, then $a > c$.
2. True or false: Absolute value equations always have two solutions.

Question 3 refers to the following inequality.

$$7x + 3 \leq 9x - 7$$

3. Solve the inequality above. Write the solution in interval notation and graph it on a number line.
4. Graph: $\{x \mid x > -2\}$.
5. Which of the following inequalities is not equivalent to $x + 4 \leq 7$?
   - **A.** $2x + 3 \leq 9$
   - **B.** $4x + 1 \leq 6x - 5$
   - **C.** $-\frac{x}{3} + 4 \geq 3$
   - **D.** $-3x \geq -9$

Question 6 refers to the following graph.

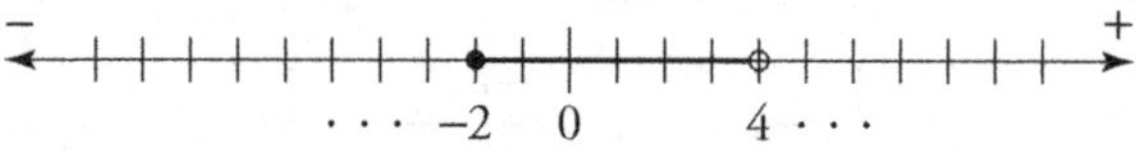

6. Which of the following inequalities matches the graph above?
   - **A.** $-2 \leq x < 4$
   - **B.** $-2 < x$ and $x \leq 4$
   - **C.** $-2 \geq x > 4$
   - **D.** $x > -2$ and $x \leq 4$
7. Simplify $3 - |-5|$.

8. Solve for x: $4|x - 1| - 3 = 17$.
9. Explain why the absolute value equation $|x| + 8 = 3$ has no solution.
10. Write an absolute value equation that has 4 and -2 as solutions.
11. Solve and graph $2|x| + 4 > 8$.
12. Solve and graph $5|x - 2| + 8 \leq 33$.

Answers

1. True
2. False
3. The solution in interval notation is $[5, \infty)$, and the graph on a number line is

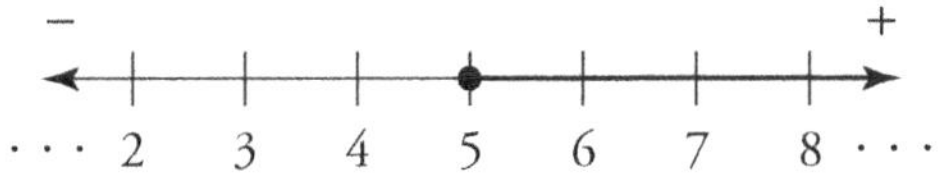

4.

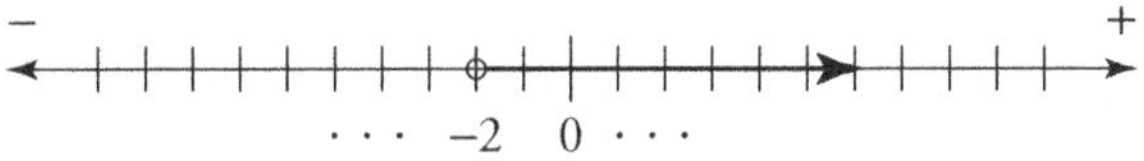

5. B
6. A
7. -2
8. $-4, 6$
9. When isolating the absolute value expression, you get $|x| = -5$. Since distance can't be negative, there are no values of x that are negative 5 units from zero.
10. Answers may vary. One possible solution is $|x - 1| = 3$.
11. $x > 2$ or $x < -2$

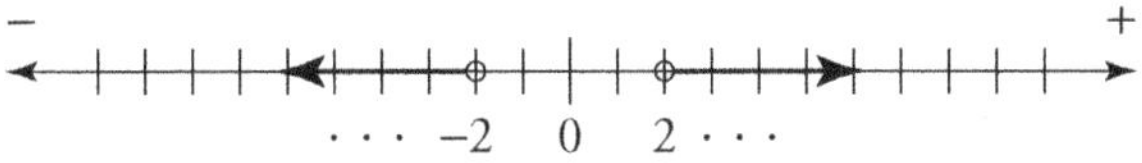

12. $-3 \leq x \leq 7$

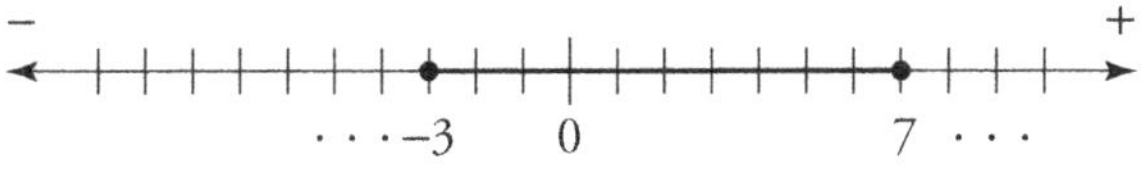

Chapter 8
COORDINATE GEOMETRY

Chapter Check-In

- ❑ The coordinate plane
- ❑ Graphing equations on a coordinate plane
- ❑ Slope and intercept of linear equations
- ❑ Graphing linear equations using slope and intercept
- ❑ Writing linear equations
- ❑ Linear inequalities and half-planes

Common Core Mathematics uses coordinates to reason and prove simple geometric theorems algebraically. Coordinate geometry is a system of geometry that deals with graphing (or plotting) and analyzing points, lines, and areas on the coordinate plane (coordinate graph). In this chapter, you will see algebra and geometry interconnected through coordinate graphs of lines and curves.

Common Core Standard: Creating Equations and Reasoning with Equations and Inequalities

Create equations in two or more variables to represent relationships between quantities, and graph equations on coordinate axes with labels and scales (A.CED.2). Rearrange formulas to highlight a quantity of interest, using the same reasoning as in solving equations (A.CED.4). Understand that the graph of an equation in two variables is the set of all its solutions plotted in the coordinate plane, often forming a curve (which could be a line) (A.REI.10).

Common Core Standard: Interpreting Functions

Calculate and interpret the average rate of change of a function (presented symbolically or as a table) over a specified interval. Estimate the rate of change from a graph (F.IF.6). Graph functions expressed symbolically and show key features of the graph, by hand in simple cases and using technology for more complicated cases (F.IF.7).

Common Core Standard: Interpreting Categorical and Quantitative Data

Interpret the slope (rate of change) and the intercept (constant term) of a linear model in the context of the data (S.ID.7).

The Coordinate Plane

Common Core Mathematics asks that you understand and graph points on a coordinate plane to solve real-world and mathematical problems. Each point on a number line is assigned a number. In the same way, each point on a plane is assigned a pair of numbers. These numbers represent the placement of the point relative to two intersecting number lines. In the **coordinate plane** (see figure below), two perpendicular number lines are used and are called **coordinate axes.** One axis is horizontal and is called the ***x*-axis.** The other is vertical and is called the ***y*-axis.** The point of intersection of the two number lines is called the **origin** and is represented by the coordinates (0, 0).

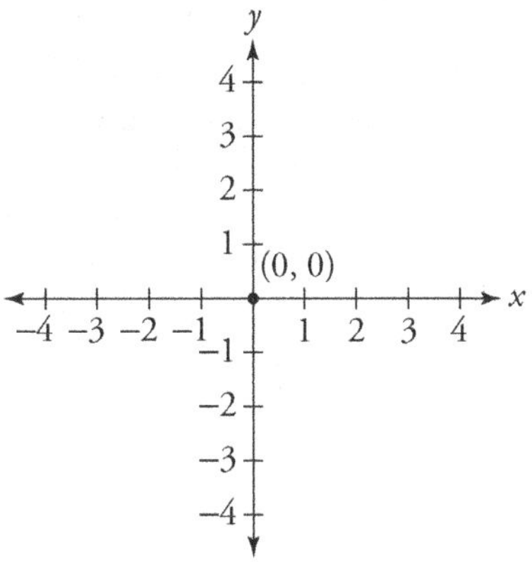

The location of each point on a plane is named by a unique **ordered pair** of numbers called the **coordinates,** (x, y). Some coordinates are noted in the following figure.

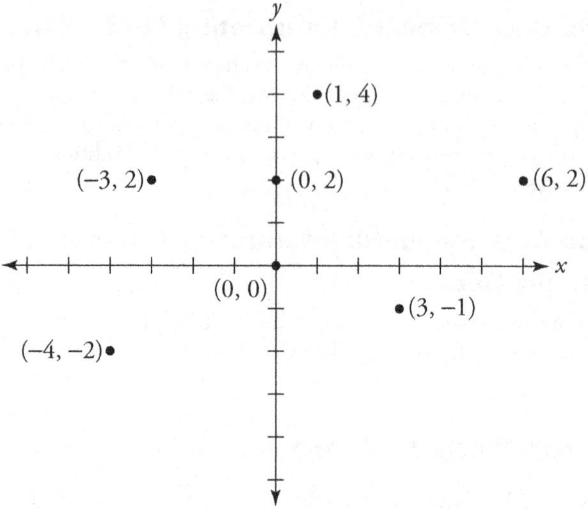

Notice that on the x-axis, numbers to the right of 0 are positive and numbers to the left of 0 are negative. On the y-axis, numbers above 0 are positive and numbers below 0 are negative. Also, note that the first number in the ordered pair is called the ***x*-coordinate,** or **abscissa,** and the second number is the ***y*-coordinate,** or **ordinate.** The x-coordinate shows the right or left direction, and the y-coordinate shows the up or down direction. To reach the point (x, y), first use the x-coordinate to move to the right (positive) or the left (negative) from the place of origin. Then, use the y-coordinate to move up (positive) or down (negative).

The coordinate plane is divided into four regions called **quadrants.** Common Core Mathematics will often denote real-world and mathematical problems by graphing in quadrants of a coordinate plane, and interpret coordinate values of points in the context of the situation. The following figure shows the quadrants of the coordinate plane.

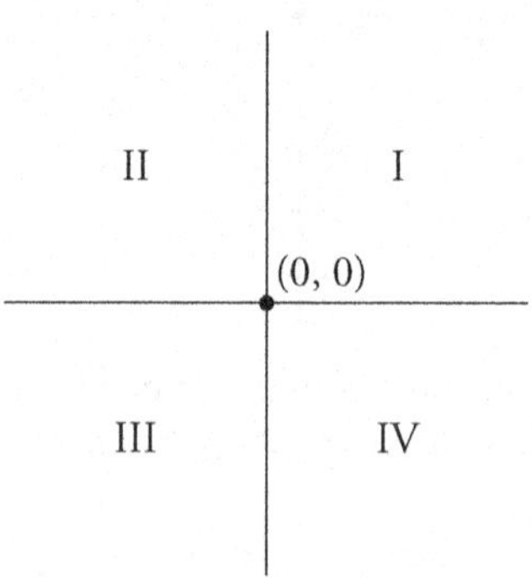

Notice the following:

In quadrant I, x and y are both always positive.

In quadrant II, x is always negative and y is always positive.

In quadrant III, x and y are both always negative.

In quadrant IV, x is always positive and y is always negative.

Graphing equations on the coordinate plane

To graph an equation in two variables, x and y, find a coordinate by giving a value to one variable and solving the resulting equation for the other value. Repeat this process to find other coordinates. (When giving a value for one variable, you could start with 0, then try 1, and so on.) Then graph the equation by plotting enough coordinates to recognize a pattern. Next, connect the coordinates and extend the graph. The solution of an equation in two variables is the set of all (x, y) points that will make the equation true. The set of all its solutions is plotted on the coordinate plane; the graph of an equation often forms a curve (which could be a line).

Example 1: Graph the equation $x + y = 6$.

If $x = 0$, then $y = 6$.

$$\begin{aligned} (0) + y &= 6 \\ y &= 6 \end{aligned}$$

If $x = 1$, then $y = 5$.

$$\begin{aligned} (1) + y &= 6 \\ -1 \quad & -1 \\ \hline y &= 5 \end{aligned}$$

If $x = 2$, then $y = 4$

$$\begin{aligned} (2) + y &= 6 \\ -2 \quad & -2 \\ \hline y &= 4 \end{aligned}$$

Using a simple chart is helpful.

x	*y*	*(x, y)*
0	6	(0, 6)
1	5	(1, 5)
2	4	(2, 4)

Now plot these coordinates as shown in the following graph.

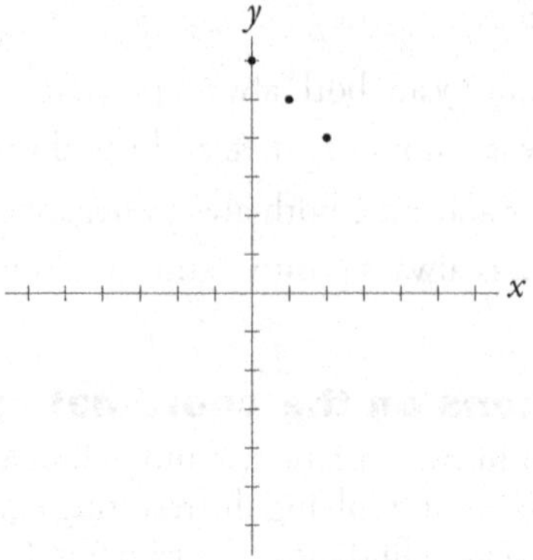

Notice that these solutions, when plotted, form a straight line. Equations whose solution sets form a straight line are called **linear equations.** Complete the graph of $x + y = 6$ by drawing the line that passes through these points, as shown in the following figure.

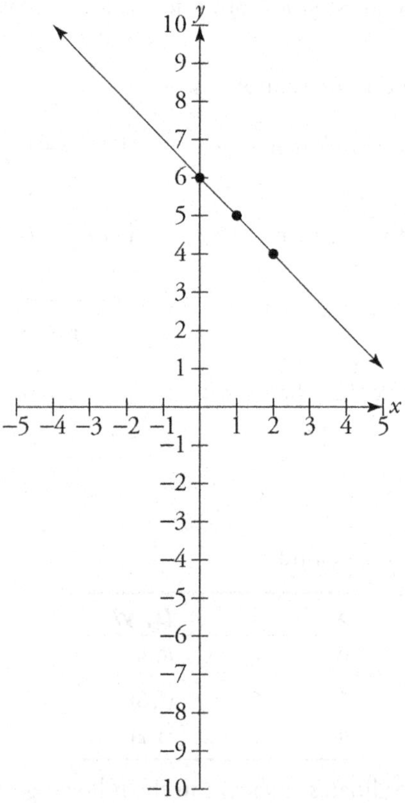

Equations that have a variable raised to a power, show division by a variable, involve variables with square roots, or have variables multiplied together will not form a straight line when their solutions are graphed. These are called **non-linear equations.**

Example 2: Graph the equation $y = x^2 + 4$.

If $x = 0$, then $y = 4$.

$$y = (0)^2 + 4$$
$$y = 0 + 4$$
$$y = 4$$

If $x = 1$ or -1, then $y = 5$.

$$y = (1)^2 + 4 \qquad y = (-1)^2 + 4$$
$$y = 1 + 4 \qquad y = 1 + 4$$
$$y = 5 \qquad y = 5$$

If $x = 2$ or -2, then $y = 8$.

$$y = (2)^2 + 4 \qquad y = (-2)^2 + 4$$
$$y = 4 + 4 \qquad y = 4 + 4$$
$$y = 8 \qquad y = 8$$

Use a simple chart.

x	*y*	*(x, y)*
−2	8	(−2, 8)
−1	5	(−1, 5)
0	4	(0, 4)
1	5	(1, 5)
2	8	(2, 8)

Now plot these coordinates as shown in the figure below.

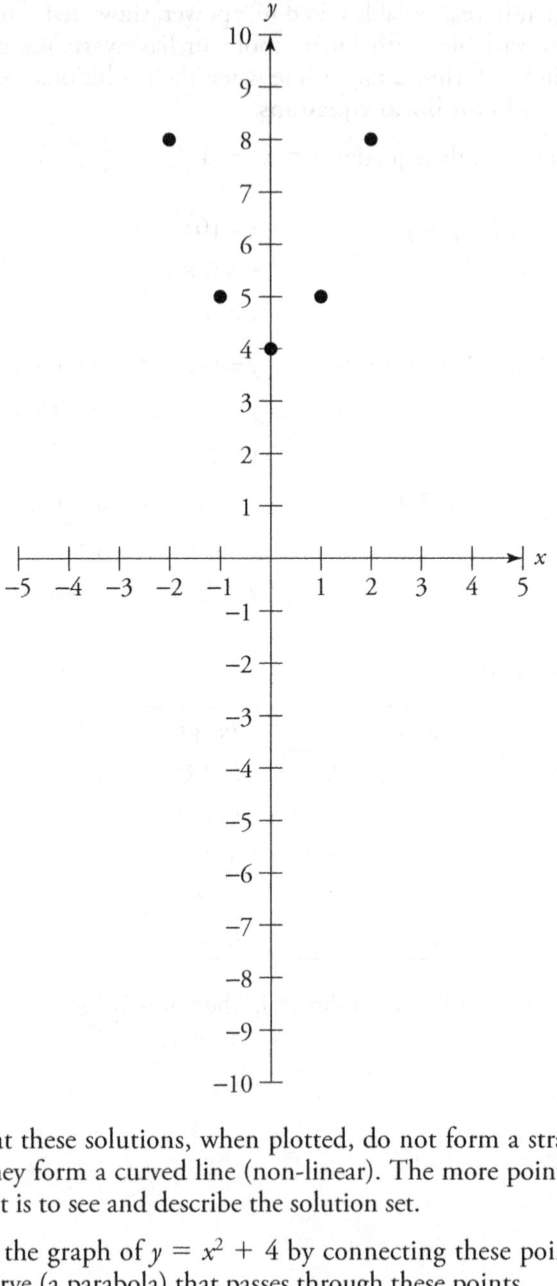

Notice that these solutions, when plotted, do not form a straight line. Instead, they form a curved line (non-linear). The more points plotted, the easier it is to see and describe the solution set.

Complete the graph of $y = x^2 + 4$ by connecting these points with a smooth curve (a parabola) that passes through these points.

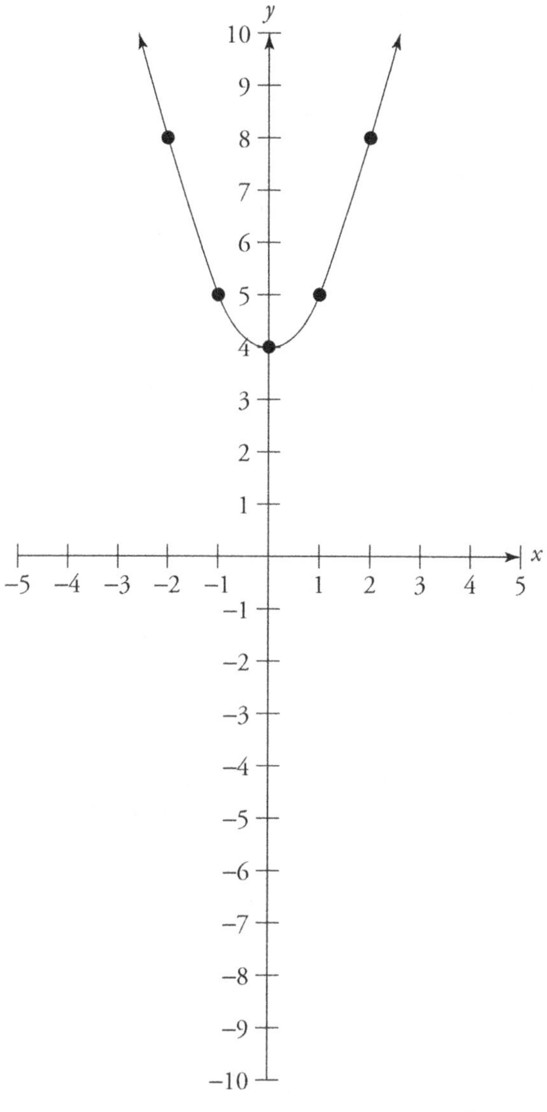

Slope and Intercept of Linear Equations

There are two relationships between the graph of a linear equation and the equation itself that must be pointed out. One involves the slope of the line, or constant *rate of change,* and the other involves the point where the line crosses the y-axis, called the **y-intercept.** In order to see either of these relationships, the terms of the equation must be in a certain order.

$$(+)(1)y = (\quad)x + (\quad)$$

When the terms are written in this order, the equation is said to be in slope-intercept form. **Slope-intercept form** is written $\boldsymbol{y = mx + b}$, and the two relationships involve m, the slope, and b, the y-intercept. The slope (m) is the numerical coefficient of the x term and the y-intercept (b) is the constant. To write a linear equation in this form, you must solve the equation for y.

Example 3: Write the linear equations in slope-intercept form.

(a) $x - y = 3$

$$-y = -x + 3$$
$$y = x - 3$$

(b) $y = -2x + 1$ (already in slope-intercept form)

(c) $x - 2y = 4$

$$-2y = -x + 4$$
$$2y = x - 4$$
$$y = \frac{1}{2}x - 2$$

As shown in the graphs of the three problems in Example 3, the lines cross the y-axis at -3, $+1$, and -2, the constant in each equation.

Two points in the coordinate plane can be used to find the **slope of a line.** The slope is defined as the ratio of the vertical change to the horizontal change.

$$\text{slope} = m = \frac{\text{the change in } y}{\text{the change in } x} = \frac{\Delta y}{\Delta x} = \frac{\text{rise}}{\text{run}}$$

The word "change" refers to the difference in the value of y (or x) between two points on the line.

$$\text{slope of line } AB = \frac{y_A - y_B}{x_A - x_B}\left[\frac{y \text{ at point } A - y \text{ at point } B}{x \text{ at point } A - x \text{ at point } B}\right]$$

Note: Points A and B can be any two points on a line; there will be no difference in the slope.

Example 4: Find the slope of $x - y = 3$ shown in the graph below using coordinates.

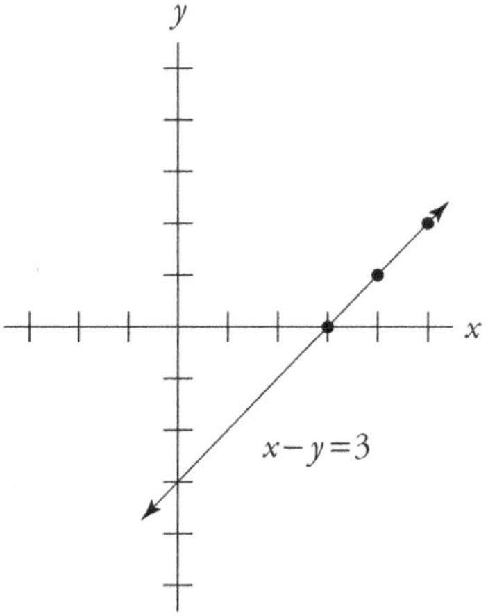

To find the slope of the line, pick any two points on the line, such as *A* (3, 0) and *B* (5, 2), and calculate the slope.

$$\text{slope } = \frac{y_A - y_B}{x_A - x_B} = \frac{(0)-(2)}{(3)-(5)} = \frac{-2}{-2} = 1$$

Example 5: Find the slope of $y = -2x + 1$ using coordinates.

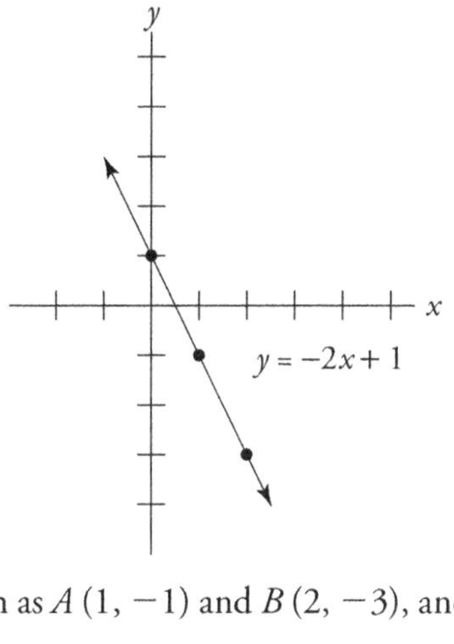

Pick two points, such as *A* (1, −1) and *B* (2, −3), and calculate the slope.

$$\text{slope} = \frac{y_A - y_B}{x_A - x_B} = \frac{-1-(-3)}{1-2} = \frac{-1+3}{1-2} = \frac{2}{-1} = -2$$

Example 6: Find the slope of $x - 2y = 4$ using coordinates.

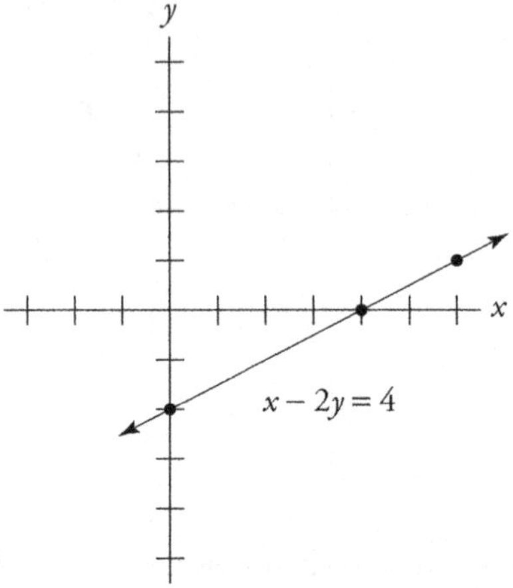

Pick two points, such as A (0, −2) and B (4, 0), and calculate the slope.

$$\text{slope} = \frac{y_A - y_B}{x_A - x_B} = \frac{(-2)-(0)}{(0)-(4)} = \frac{-2}{-4} = \frac{1}{2}$$

You may have noticed that the three equations in Example 3 are equivalent to those in Examples 4, 5, and 6. However, in Example 3, they are written in slope-intercept form. As you can see, when an equation is written in slope-intercept form, the slope of the line is the same as the numerical coefficient of the x term.

(a) $y = 1x - 3$

slope = 1 y-intercept = −3

(b) $y = -2x + 1$

slope = −2 y-intercept = 1

(c) $y = \frac{1}{2}x - 2$

slope $= \frac{1}{2}$ y-intercept = −2

Example 7: Find the slope of a line using the table below.

x	*y*
−2	8
−1	6
0	4
1	2
2	0

Choose any two ordered pairs in the table and use the slope formula to find the slope of the line. Let's choose (−2, 8) and (1, 2).

$$\text{slope} = \frac{y_A - y_B}{x_A - x_B} = \frac{(8)-(2)}{(-2)-(1)} = \frac{6}{-3} = -2$$

Vertical and horizontal lines

Equations of lines can take many different forms. Vertical and horizontal lines have equations with only one variable. Equations of vertical lines are in the form $x = h$ and horizontal lines are in the form $y = k$, where h and k are constants.

The equation $x = 2$ represents a vertical line that crosses at the x-axis at (2, 0), as shown below.

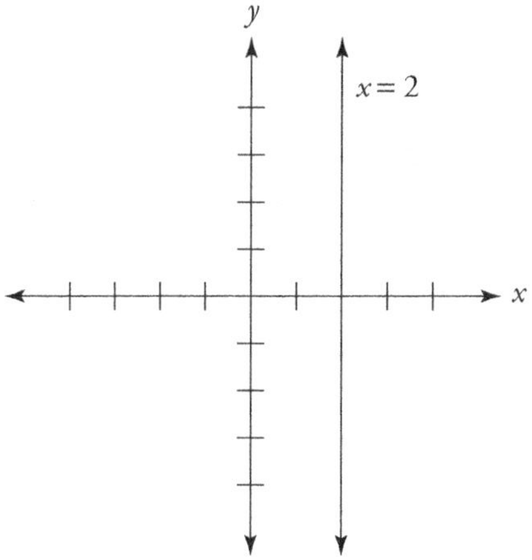

Recall that the x-intercept of a line is the point where the graph crosses the x-axis. The x-intercept of the line in the figure above is 2. Except for the vertical line $x = 0$ (the y-axis), a vertical line does not have a y-intercept.

The equation $y = 5$ represents a horizontal line that crosses the y-axis at (0, 5), as shown below.

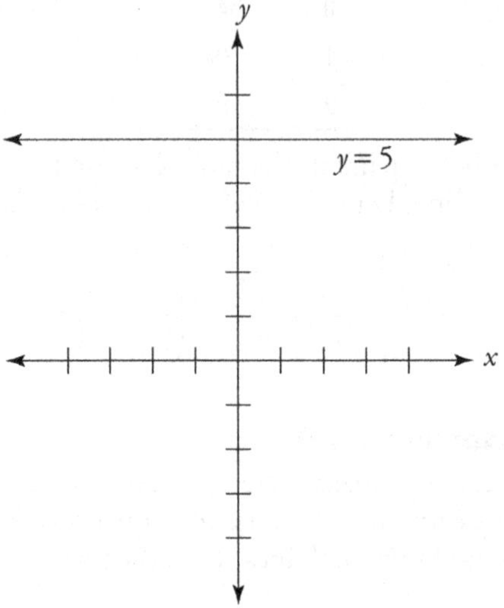

The y-intercept of the line shown above is 5. Except for the horizontal line $y = 0$ (the x-axis), a horizontal line does not have an x-intercept.

The slopes of these lines are different from lines in the form $y = mx + b$. A horizontal line has a slope of zero. A vertical line has no slope or an undefined slope.

Graphing linear equations using slope and intercept

Graphing a linear equation by using its slope and y-intercept is usually quite easy.

1. Write the equation in slope-intercept form.
2. Locate the y-intercept on the graph (that is, find one of the points on the line).
3. Write the slope as a ratio (fraction), and use it to locate other points on the line using the concept of $\frac{\text{rise}}{\text{run}}$, where rise means you go up or down depending on the sign of the number, and run means you go left or right depending on the sign of the number.
4. Draw a straight line through the coordinates.

Example 8: Graph the equation $x - y = 2$ using slope-intercept form.

$$x - y = 2$$
$$-y = -x + 2$$
$$y = x - 2$$

Locate -2 on the y-axis and from this point count as shown below; slope = 1.

$$\text{or } \frac{1}{1} \begin{matrix}\text{(for every 1 up)} \\ \text{(go 1 to the right)}\end{matrix}$$

$$\text{or } \frac{-1}{-1} \begin{matrix}\text{(for every 1 down)} \\ \text{(go 1 to the left)}\end{matrix}$$

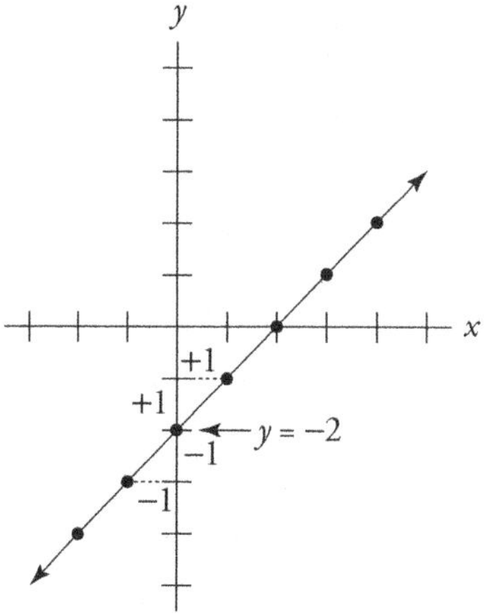

Example 9: Graph the equation $2x - y = -4$ using slope-intercept form.

$$2x - y = -4$$
$$-y = -2x - 4$$
$$y = 2x + 4$$

Locate 4 on the y-axis and from this point count as shown; slope = 2.

$$\text{or } \frac{2}{1} \begin{matrix}\text{(for every 2 up)} \\ \text{(go 1 to the right)}\end{matrix}$$

$$\text{or } \frac{-2}{-1} \begin{matrix}\text{(for every 2 down)} \\ \text{(go 1 to the left)}\end{matrix}$$

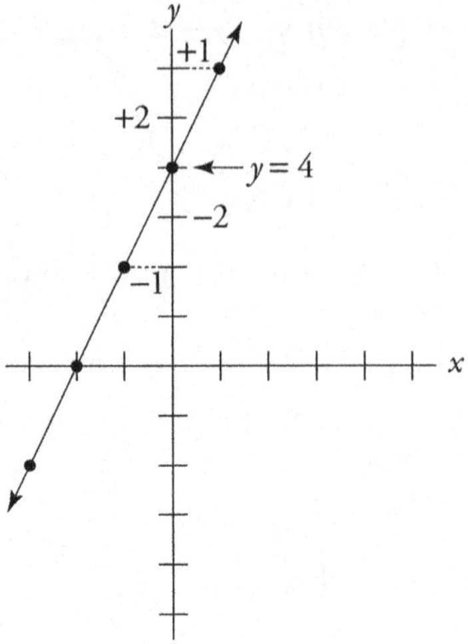

Example 10: Graph the equation $x + 3y = 0$ using slope-intercept form.

$$x+3y=0$$
$$3y=-x+(0)$$
$$y=-\frac{1}{3}x+(0)$$

Locate 0 on the y-axis and from this point count as shown; slope $= -\frac{1}{3}$.

or $\frac{-1}{3}$ (for every 1 down) (go 3 to the right)

or $\frac{1}{-3}$ (for every 1 up) (go 3 to the left)

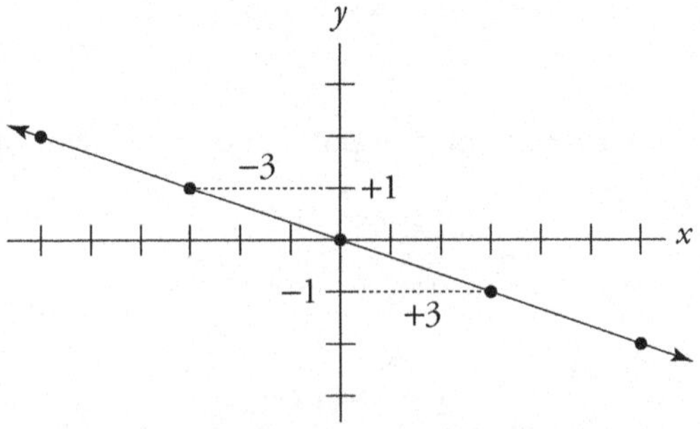

Writing linear equations

Common Core Mathematics asks you to represent and write real-world and mathematical problems graphically. To write a linear equation when working with ordered pairs, slopes, and intercepts, use one of the following approaches, depending on which form of the equation you want to have. The three most common are the **slope-intercept form,** the **point-slope form,** and the **standard form of a linear equation in two variables.**

As we've discussed, the slope-intercept form looks like $y = mx + b$, where m is the slope of the line and b is the y-intercept. The point-slope form looks like $y - y_1 = m(x - x_1)$, where m is the slope of the line and (x_1, y_1) is any point on the line. The standard form looks like $Ax + By = C$, where, if possible, A, B, and C are integers and A and B are not both zero.

Method 1. Slope-intercept form:

1. Find the slope, m.
2. Find the y-intercept, b.
3. Substitute the slope and y-intercept into the slope-intercept form, $y = mx + b$.

Note: Slope-intercept form is a useful and quick way to graph linear equations.

Method 2. Point-slope form:

1. Find the slope, m.
2. Use any point known to be on the line.
3. Substitute the slope and the ordered pair of the point into the point-slope form, $y - y_1 = m(x - x_1)$.

Note: You could begin with the point-slope form for the equation of the line and then solve the equation for y. You will get the slope-intercept form without having to first find the y-intercept.

Method 3. Standard form:

1. Find the equation of the line using either the slope-intercept form or the point-slope form.
2. With appropriate algebra, rewrite or transform the equation to get the x terms and the y terms on one side of the equation and the constant on the other side of the equation.
3. If necessary, multiply each side of the equation by the least common denominator to have all integer coefficients for the variables.

Note: Standard form can be used to quickly find x- and y-intercepts of a linear equation and to graph using the intercepts.

Example 11: Graph the linear function $2x - 3y = -12$ using intercepts.

To algebraically find the y-intercept of a linear function, set $x = 0$ and solve for y. To find the x-intercept of a linear function, set $y = 0$ and solve for x.

$$\begin{aligned} x &= 0 \\ 2(0) - 3y &= -12 \\ -3y &= -12 \\ y &= 4 \text{ (y-intercept)} \end{aligned} \qquad \begin{aligned} y &= 0 \\ 2x - 3(0) &= -12 \\ 2x &= -12 \\ x &= -6 \text{ (x-intercept)} \end{aligned}$$

We will use the coordinates $(0, 4)$ and $(-6, 0)$ to graph the linear function.

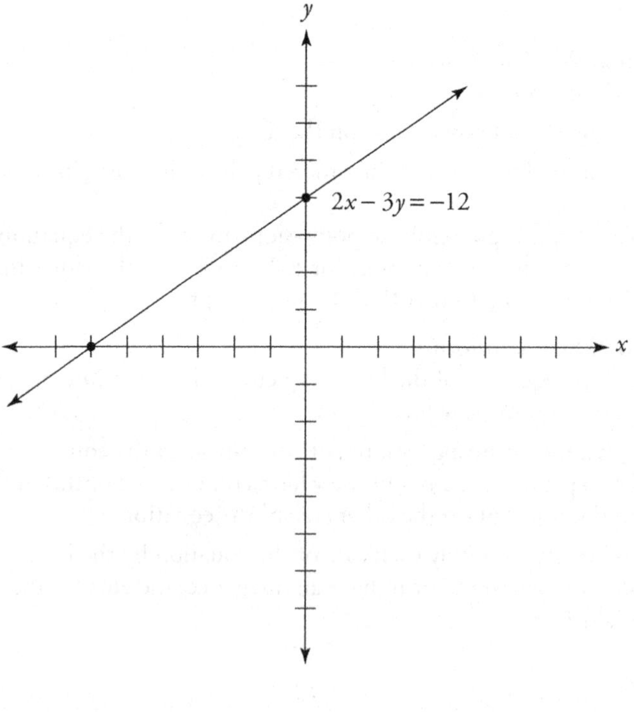

Example 12: Write the equation of the line in slope-intercept form, where $m = -4$ and $b = 3$. Then convert it into standard form.

1. Find the slope, m.

$$m = -4 \text{ (given)}$$

2. Find the y-intercept, b.

$$b = 3 \text{ (given)}$$

3. Substitute the slope and y-intercept into the equation, $y = mx + b$.

$$y = -4x + 3 \text{ (slope-intercept form)}$$

4. With appropriate algebra, rewrite the equation to get the x terms and the y terms on one side of the equation and the constant on the other side of the equation.

$$\begin{array}{rcl} & y = -4x + 3 & \\ +4x & \quad +4x & \\ \hline 4x + y = & \quad 3 & \text{(standard form)} \end{array}$$

Example 13: Find the equation of the line, in point-slope form, passing through the point (6, 4) with a slope of -3. Then rewrite the equation in standard form.

1. Find the slope, m.

$$m = -3 \text{ (given)}$$

2. Use any point known to be on the line.

$$(6, 4) \text{ (given)}$$

3. Substitute the slope and the ordered pair of the point into the point-slope form.

$$y - y_1 = m(x - x_1)$$
$$y - 4 = -3(x - 6) \quad \text{(point-slope form)}$$

4. With appropriate algebra, rewrite the equation to get the x terms and the y terms on one side of the equation and the constant on the other side of the equation.

$$\begin{array}{rcl} y-4 &=& -3(x-6) \\ y-4 &=& -3x+18 \\ +3x & & +3x \\ \hline 3x+y-4 &=& 18 \\ +4 & & +4 \\ \hline 3x+y &=& 22 \end{array} \quad \text{(standard form)}$$

Example 14: Find the equation of the line, in either slope-intercept form or point-slope form, passing through (5, −4) and (3, 7). Then convert it to standard form.

Method 1. Slope-intercept form:

1. Find the slope, m.

$$m = \frac{\text{change in } y}{\text{change in } x}$$

$$m = \frac{(-4)-7}{5-3} = \frac{-11}{2} = -\frac{11}{2}$$

2. Find the y-intercept, b.

   Substitute the slope and either point into the equation.

$$y = mx + b, \text{ where } m = -\frac{11}{2},\ x = 5,\ y = -4$$

$$\begin{array}{rcl} -4 &=& -\frac{11}{2}(5)+b \\ -4 &=& -\frac{55}{2}+b \\ +\frac{55}{2} & & +\frac{55}{2} \\ \hline \frac{47}{2} &=& b \end{array}$$

3. Substitute the slope and y-intercept into the equation, $y = mx + b$.

$$\text{Since} \quad m = -\frac{11}{2}$$

$$\text{and} \quad b = \frac{47}{2}$$

$$\text{then} \quad y = -\frac{11}{2}x + \frac{47}{2} \quad \text{(slope-intercept form)}$$

4. With appropriate algebra, rewrite the equation to get the x terms and the y terms on one side of the equation and the constant on the other side of the equation. If necessary, multiply each side of the equation by the least common denominator to have all integer coefficients for the variables.

$$\begin{array}{rcl} & y = -\frac{11}{2}x + \frac{47}{2} & \\ +\frac{11}{2}x & & +\frac{11}{2}x \\ \hline \frac{11}{2}x + y & = & \frac{47}{2} \end{array}$$

$$2\left(\frac{11}{2}x + y\right) = 2\left(\frac{47}{2}\right)$$

$$11x + 2y = 47 \quad \text{(standard form)}$$

Method 2. Point-slope form:

1. Find the slope, m.

$$m = \frac{\text{change in } y}{\text{change in } x}$$

$$m = \frac{(-4) - 7}{5 - 3} = \frac{-11}{2} = -\frac{11}{2}$$

2. Use any point known to be on the line.

(3, 7) (given)

3. Substitute the slope and the ordered pair of the point into the equation.

$$y - y_1 = m(x - x_1)$$

$$y - 7 = -\frac{11}{2}(x - 3) \qquad \text{(point-slope form)}$$

4. With appropriate algebra, rewrite the equation to get the x terms and the y terms on one side of the equation and the constant on the other side of the equation.

   If necessary, multiply each side of the equation by the least common denominator to have all integer coefficients for the variables.

$$y - 7 = -\frac{11}{2}(x - 3)$$

$$y - 7 = -\frac{11}{2}x + \frac{33}{2}$$

$$+\frac{11}{2}x \qquad\qquad +\frac{11}{2}x$$

$$\frac{11}{2}x + y - 7 = \frac{33}{2}$$

$$+7 \qquad\qquad +7$$

$$\frac{11}{2}x + y \quad = \frac{47}{2}$$

$$2\left(\frac{11}{2}x + y\right) = 2\left(\frac{47}{2}\right)$$

$$11x + 2y = 47 \qquad \text{(standard form)}$$

Linear Inequalities and Half-Planes

Common Core Standard: Creating Equations and Reasoning with Equations and Inequalities

Represent constraints by equations or inequalities, and by systems of equations and/or inequalities, and interpret solutions as viable or non-viable options in a modeling context (A.CED.3). Graph the solutions to a linear inequality in two variables as a half-plane (excluding the boundary in the case of a strict inequality), and graph the solution set to a system of linear inequalities in two variables as the intersection of the corresponding half-planes (A.REI.12).

Each line plotted on the coordinate plane divides the graph (or plane) into two **half-planes.** This line is called the *boundary line* (or *bounding line*). Before graphing a linear inequality in two variables, you must first find or use the equation of the line to make a boundary line. The solution of a linear inequality in two variables is any ordered pair that will make the inequality a true statement. The set of all such points in the coordinate plane is called the *graph of the inequality,* which represents a shaded half-plane. The line itself may or may not be included in the solution set.

Open half-plane

If the inequality is a ">" or "<", then the graph will be an **open half-plane.** An open half-plane does not include the boundary line, so the boundary line is written as a dashed line on the graph. The dashed line indicates that the points on the line are not included in the solution set of the inequality.

Example 15: Graph the inequality $y < x - 3$.

First graph the line $y = x - 3$ to find the boundary line (use a dashed line since the inequality is "<") as shown in the graph below.

x	y
3	0
0	−3
4	1

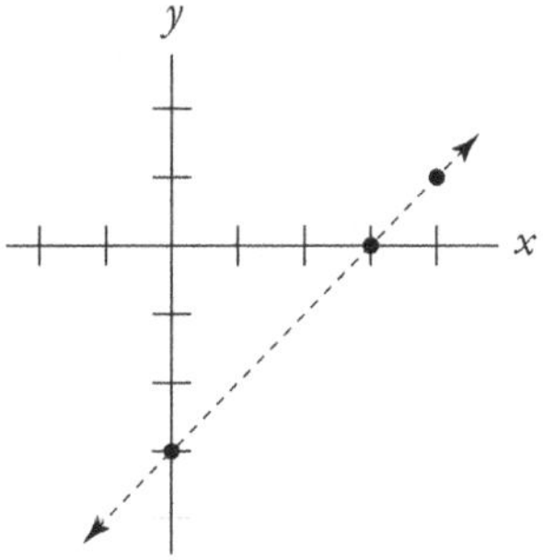

Now shade the lower half-plane since $y < x - 3$.

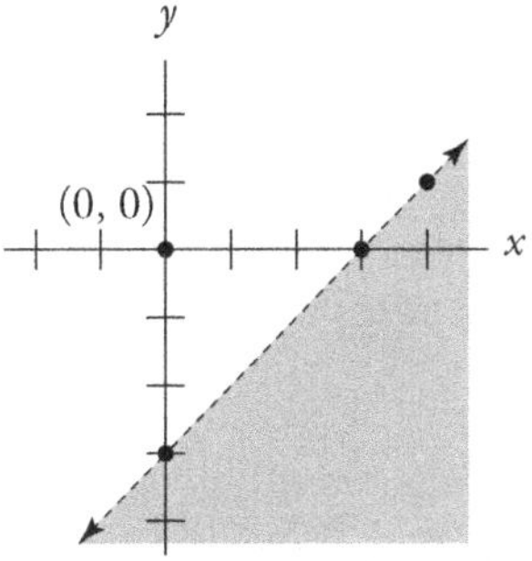

The solution region is all points below the graph of $y = x - 3$.

To check whether you've shaded the correct half-plane, substitute a pair of coordinates into the inequality—the pair of (0, 0) is often a good choice as long as the line does not go through this point. If the coordinates you selected make the inequality a true statement when substituted, then you should shade the half-plane containing those coordinates. If the coordinates you selected do not make the inequality a true statement, then shade the half-plane that does not contain those coordinates.

Since the point (0, 0) does not make this inequality a true statement,

$$y < x - 3$$

$$0 < 0 - 3 \text{ is not true.}$$

you should shade the side that does not contain the point (0, 0).

This checking method is often simply used as the method to decide which half-plane to shade.

Closed half-plane

If the inequality is a "≤" or "≥," then the graph will be a **closed half-plane.** A closed half-plane includes the boundary line and is graphed using a solid line and shading. The solid line indicates that the points on the line are included in the solution set of the inequality.

Example 16: Graph the inequality $2x - y \leq 0$.

First, write the inequality in slope-intercept form. Subtracting $2x$ from each side gives

$$-y \leq -2x$$

Now dividing each side by -1 (and changing the direction of the inequality) gives

$$y \geq 2x$$

Graph $y = 2x$ to find the boundary (use a solid line, because the inequality is "≥") as shown on the next page.

x	y
0	0
1	2
2	4

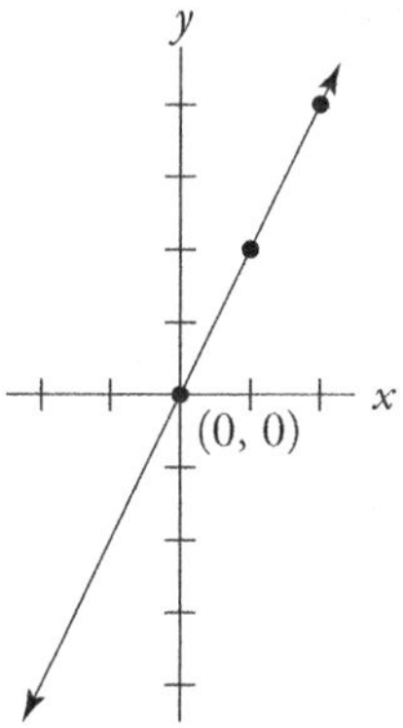

Since $y \geq 2x$, you should shade the upper half-plane. If in doubt, or to check, substitute a pair of coordinates into the inequality. Note that you do not use the point (0, 0) to decide which half-plane to shade because the line goes directly through that point. Try the pair (1, 1).

$$y \geq 2x$$
$$1 \geq 2(1)$$
$$1 \geq 2 \text{ is not true.}$$

So you should shade the half-plane that does not contain (1, 1), as shown in the graph below.

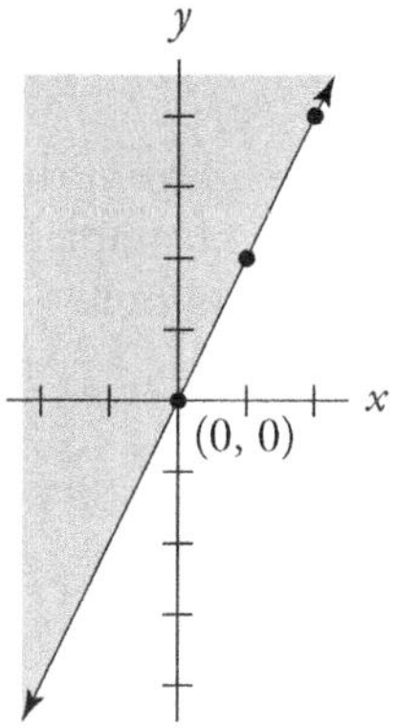

The solution region is all points on or above the graph $y = 2x$.

Chapter Check-Out

Questions

1. Is $x^2 - 8 = y$ linear or non-linear? Explain your answer.
2. Graph: $2x + y = 6$ and $2x - y = -6$ on the same coordinate plane. How are the graphs similar? How are the graphs different?
3. Find the slope and the y-intercept of the equation $Ax + By = C$.
4. Using the graph below, if the slope is doubled and the y-intercept remains the same, what is the x-intercept?

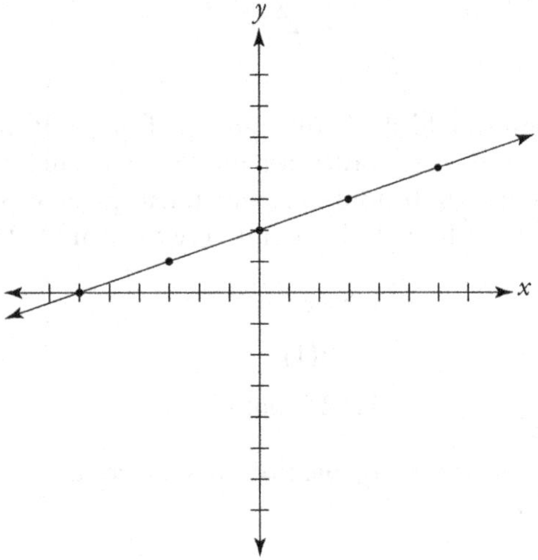

Questions 5 and 6 refer to the following information and graph.

A group of scuba divers start to ascend from the bottom of the Atlantic Ocean at 150 feet below sea level. The graph shows the elevation y (in feet) that the scuba divers ascend in x minutes.

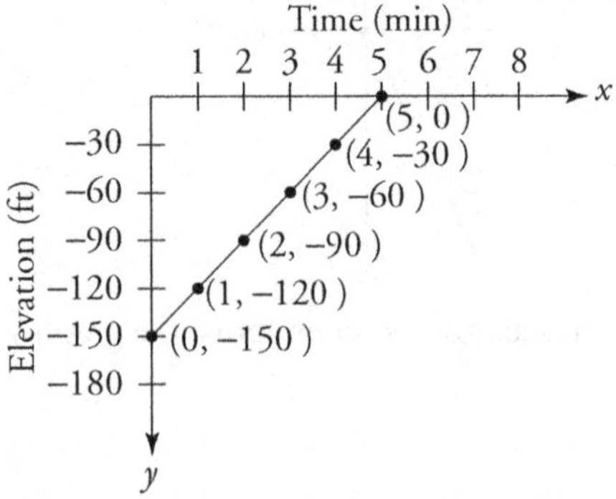

5. Find and interpret the slope of the line.
6. How many minutes does it take the scuba divers to get to the surface of the water?

Questions 7 and 8 refer to the following information.

To knit scarves and sweaters for the holidays, you buy 14 balls of yarn online to knit two sweaters at a cost of $53.86 (includes the cost for flat shipping rate). Two weeks later, you buy 10 balls of yarn with the same online company to make three scarves at a cost of $41.90 (includes the same cost for flat shipping rate).

7. Write a linear model that represents the total cost of yarn, $C(n)$, as a function of the number of balls of yarn ordered, n.
8. Interpret the slope and the y-intercept in the context of the situation.
9. The table below shows the linear relationship between the number of sides of a polygon, n, and the sum of the angle measures, S. Find and interpret the slope.

Number of sides (n)	3	4	5	6	7
Sum of angle measures (S)	180°	360°	540°	720°	980°

10. Find the equation of the line, in point-slope form, passing through the point (6, 1) with a slope of -2. Then write it in standard form.
11. Find the equation of the line, in slope-intercept form, passing through the points $(-3, 4)$ and $(2, -3)$. Then convert it to standard form.
12. If a friend graphs $y = 3x - 4$ and the line is dashed and shaded above, which inequality symbol should be used?
13. Graph: $y \leq x - 2$.

Questions 14–16 refer to the following information.

The maximum capacity a certain wedding venue can hold is 400 people. The venue can use a combination of small tables that seat 8 people and large tables that seat 10 people. The inequality $8x + 10y \leq 400$ represents the seating arrangement, where x is the number of small tables and y is the number of large tables.

14. Graph the linear inequality using standard form.
15. The coordinate (15, 8) is a solution of the inequality. Interpret this point according to the context of the seating situation.
16. Find two different combinations of tables the wedding venue can use to hold the maximum number of people, 400.

Answers

1. Non-linear because it is not in the form $Ax + By = C$.
2. Both lines have different slopes and x-intercepts, but the same y-intercept.

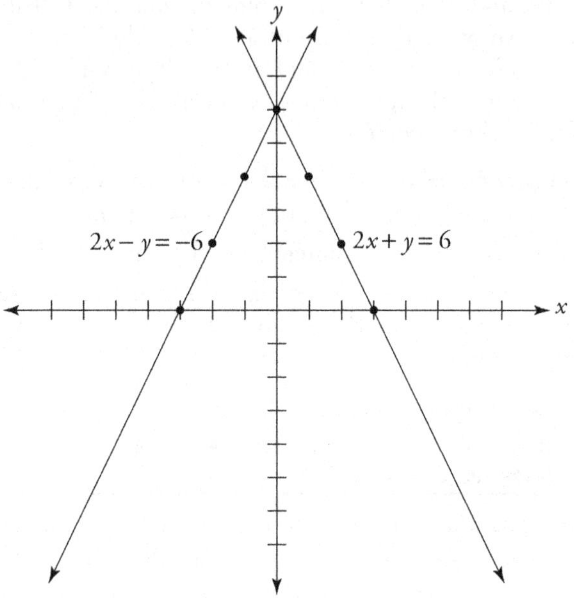

3. slope is $-\frac{A}{B}$; y-intercept is $\frac{C}{B}$
4. x-intercept is -3
5. Choosing the ordered pairs (5, 0) and (4, −30), $m = \frac{-30-0}{4-5} = \frac{-30}{-1} = 30$.

   This means that the divers ascend at a rate of 30 feet per minute.
6. 5 minutes
7. Using the points (14, 53.86) and (10, 41.90), $m = \frac{41.90-53.86}{10-14} = 2.99$, so $C(n) = 2.99x + b$. Substituting (14, 53.86) into the function, $53.86 = 2.99(14) + b$. Lastly, solving for b, $b = 12$. The linear model is $C(n) = 2.99n + 12$.

8. Since the slope is 2.99 and the *y*-intercept is 12, it costs \$2.99 for each ball of yarn and \$12 for a flat shipping rate.

9. $m = 180$. As the number of sides of a polygon increases by 1, the sum of the angle measures of the polygon increases by 180°.

10. point-slope form: $y - 1 = -2(x - 6)$; standard form: $2x + y = 13$

11. slope-intercept form: $y = -\frac{7}{5}x - \frac{1}{5}$; standard form: $7x + 5y = -1$

12. $>$

13.

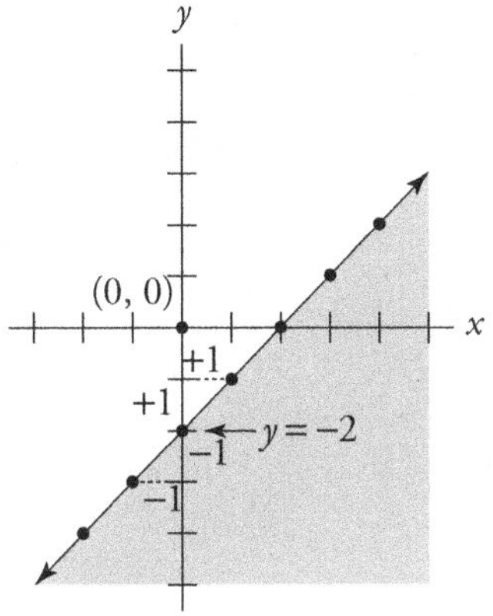

14.

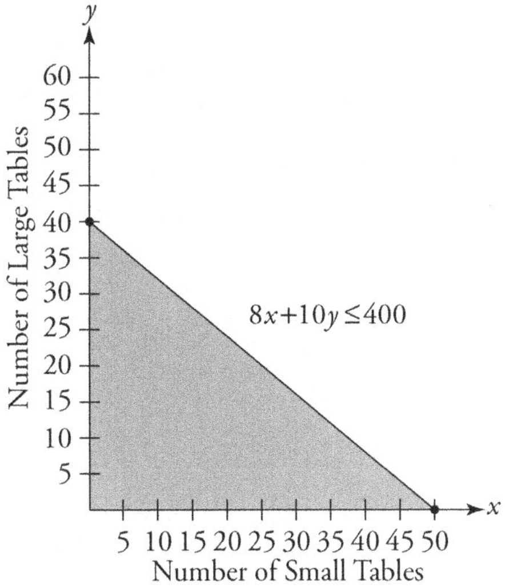

15. The possible combination of tables that could be used is 15 small tables and 8 large tables. This combination of tables would seat 200 people.
16. One combination is (50, 0), which means the wedding venue can use 50 small tables to hold the maximum capacity. The second combination is (20, 24), which means the wedding venue can use 20 small tables and 24 large tables to seat the maximum capacity.

Chapter 9

FUNCTIONS AND VARIATIONS

Chapter Check-In

- ❑ Relations and Functions
- ❑ Determining domain and range
- ❑ Evaluating functions
- ❑ Variations—direct and indirect (inverse)
- ❑ Finding the constant of variation

Functions and variations deal with relationships between two quantities, which can be represented using words, tables, graphs, and equations. A function is a relationship in which one quantity determines another in a unique way. Common Core Mathematics describes functions as involving numerical inputs and outputs that are often expressed using algebraic expressions and can model very important phenomena in the real world. For example, the return on an investment of $100 in an account earning 5% simple interest per year is a function of the length of time the money is invested.

This chapter will explore linear functions in which growth occurs at a constant rate, as demonstrated in the example above. Specific definitions and examples given in this chapter will simplify what is often unnecessarily viewed as a difficult section.

Common Core Standard: Interpreting Functions

Understand that a function from one set (called the domain) to another set (called the range) assigns to each element of the domain exactly one element of the range. If f is a function and x is an element of its domain, then $f(x)$ denotes the output of f corresponding to the input x. The graph of f is the graph of the equation $y = f(x)$ (F.IF.1). Use function notation, evaluate functions for inputs in their domains, and interpret statements that use function notation in terms of a context (F.IF.2).

Relations

Any set of ordered pairs is called a **relation,** in which input values are paired with output values. When a relation is represented using ordered pairs, the inputs are the x-coordinates and the outputs are the y-coordinates. The following is a graph of the ordered pairs $(-1, 1)$, $(1, 3)$, $(2, 2)$, $(3, 4)$.

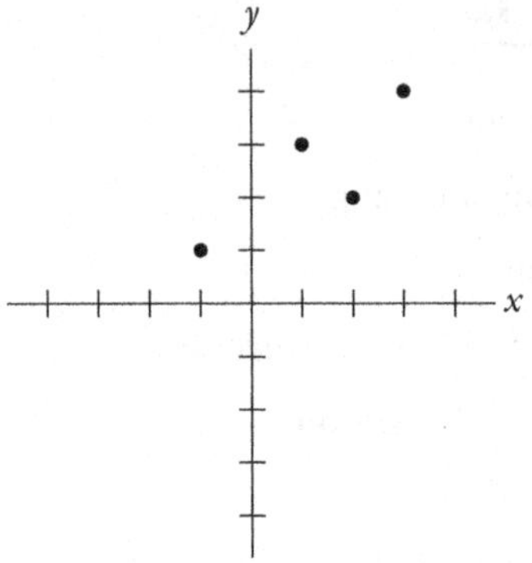

Functions

Functions are very specific types of relations. Before defining a function, it is important to define a relation.

Domain and range

The set of all x values is called the **domain** of the relation. The set of all y values is called the **range** of the relation. The domain of set A in the previous graph is $\{-1, 1, 2, 3\}$, while the range of set A is $\{1, 2, 3, 4\}$.

Example 1: Find the domain and range of the relation represented by the following graph.

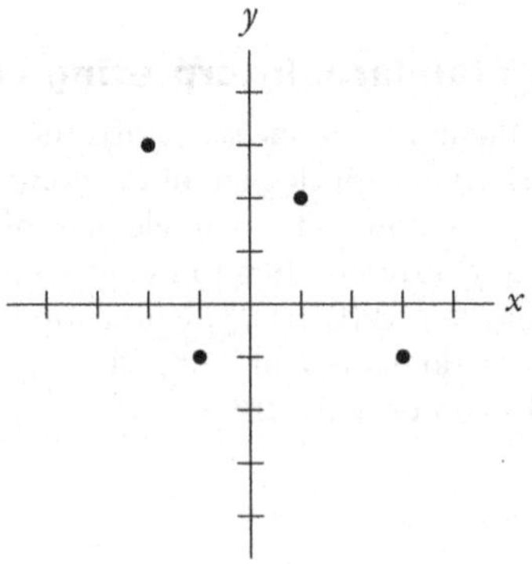

The domain is the set $\{-2, -1, 1, 3\}$. The range is the set $\{-1, 2, 3\}$.

Defining a Function

The relation in Example 1 has pairs of coordinates with unique x values. When the x value of each pair of coordinates is different, the relation is called a *function.* A **function** is a relation in which each value of the domain is paired with exactly one value of the range. A relation that pairs an x value with more than one y value is not a function. The mapping diagram below illustrates when a relation can also be considered a function.

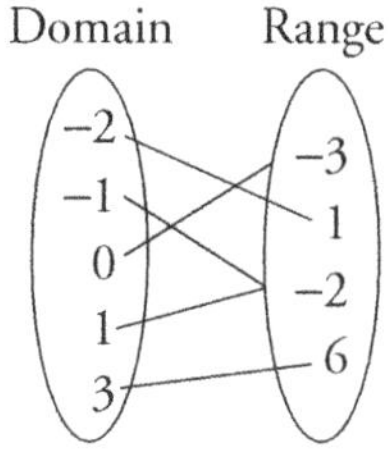

The ordered pairs from the mapping diagram above are as follows:

$$(-2, 1), (-1, -2), (0, -3), (1, -2), \text{ and } (3, 6)$$

In a function, two x values can have the same y value, but not vice versa. For example, above, both x values -1 and 1 have the y value -2: $(-1, -2)$ and $(1, -2)$. Therefore, this relation is a function.

All functions are relations, but not all relations are functions. A good example of a relation that is a function can be seen in the linear equation $y = x + 1$, graphed below. The domain and range of this function are both the set of real numbers, and the relation is a function because for any value of x there is a unique value of y.

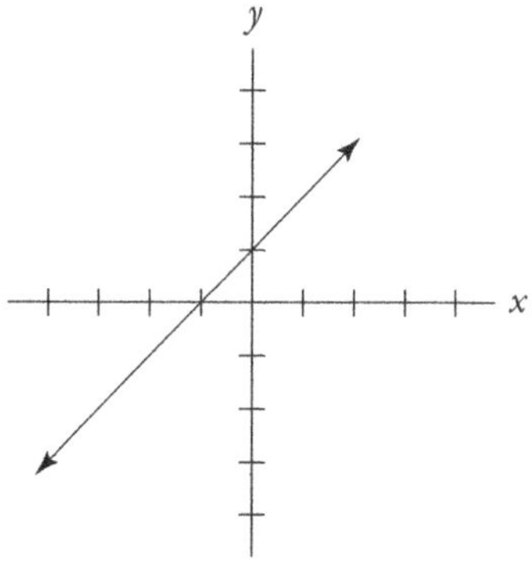

Here are a few more graphs of functions. Note that for any value of x, there is only one value for y.

Graphs of Relations that Are Functions

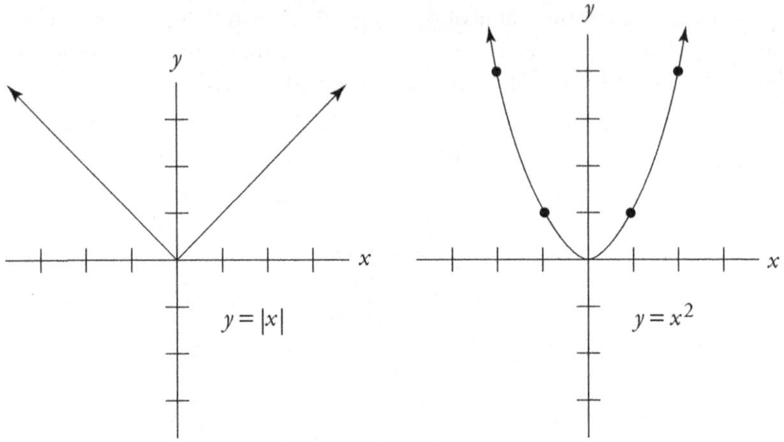

(a) (b)

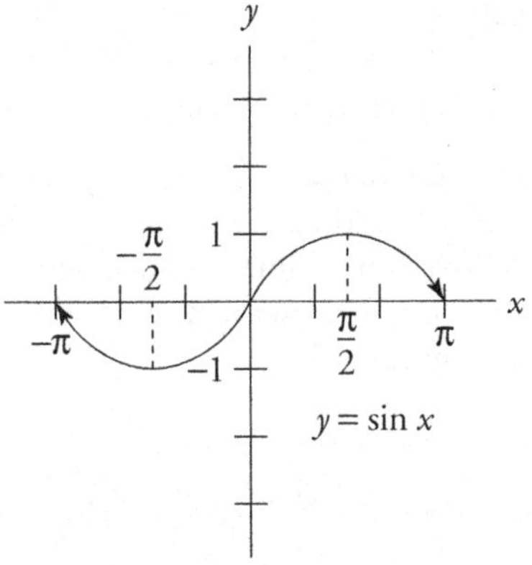

(c)

The following are graphs of relations that are not functions. In each of these relations, a single value of x is associated with two or more values of y.

Graphs of Relations that Are Not Functions

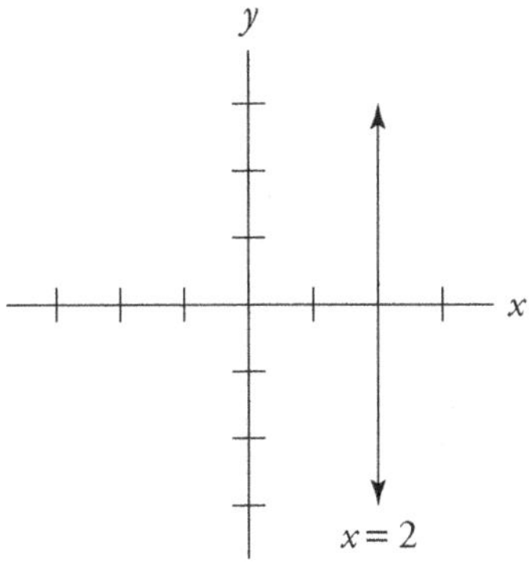

(a)

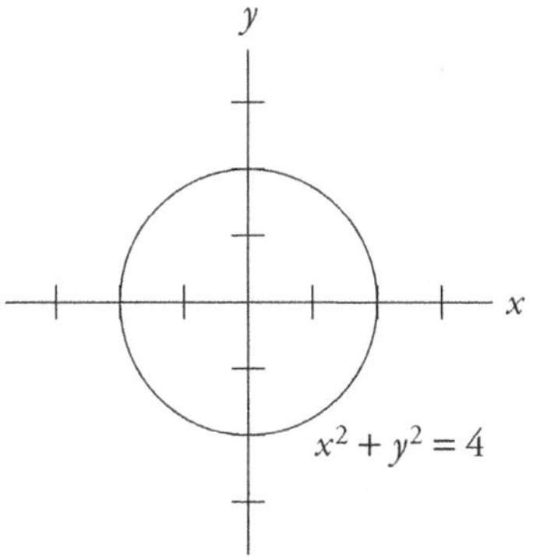

(b)

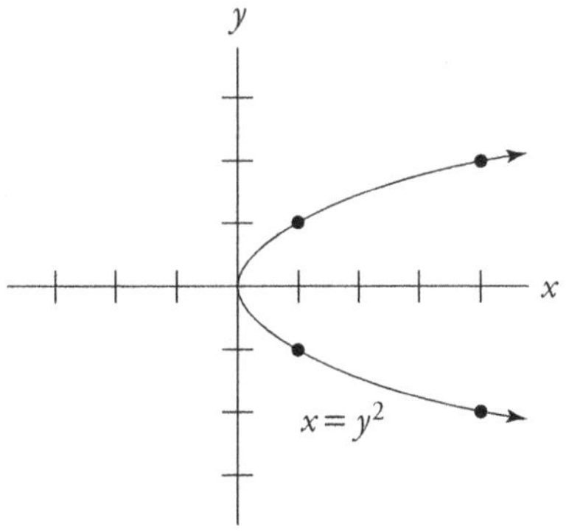

(c)

Vertical line test

The vertical line test can help you determine whether the graph of a relation is a function. If any vertical line intersects the graph in exactly one point, the relation is a function. If any vertical line intersects a graph in more than one point, the relation is not a function. For example, in the graph of $x = y^2$ above (c), if you were to draw a vertical line at $x = 4$, the line would intersect the graph in more than one point, at (4, 2) and (4, −2). Since the x value, 4, is associated with two different values of y, 2 and −2, the relation is not a function.

Determining domain, range, and if the relation is a function

Example 2: Identify the domain and range of each relation. Is the relation a function?

(a) $B = \{(-2, 3), (-1, 4), (0, 5), (1, -3)\}$ domain: $\{-2, -1, 0, 1\}$

range: $\{-3, 3, 4, 5\}$

function: yes

(b)

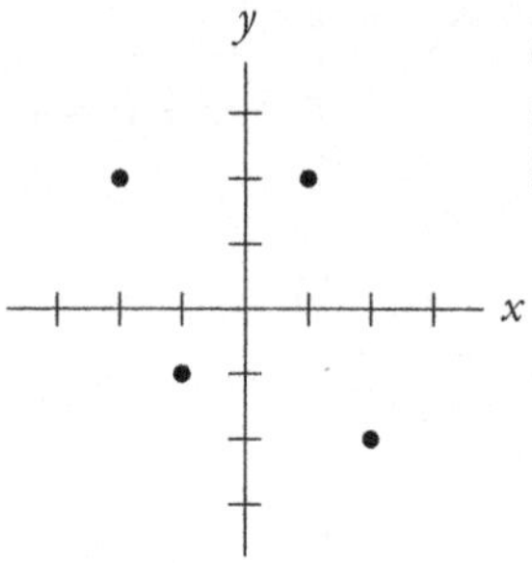

domain: $\{-2, -1, 1, 2\}$
range: $\{-2, -1, 2\}$
function: yes

(c)

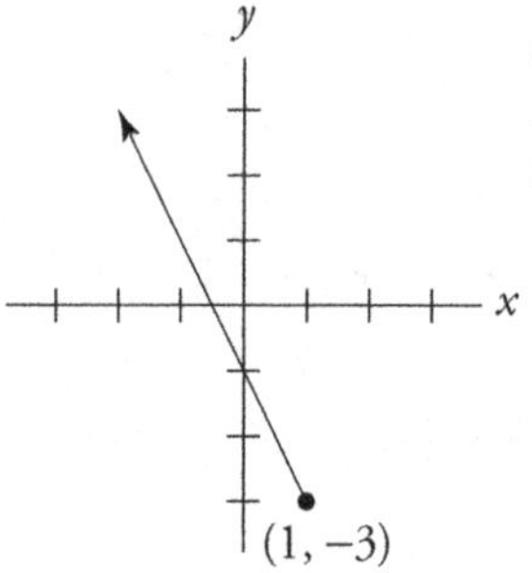

domain: $\{x | x \leq 1\}$ or $(-\infty, 1]$
range: $\{y | y \geq -3\}$ or $[-3, \infty)$
function: yes

(d)

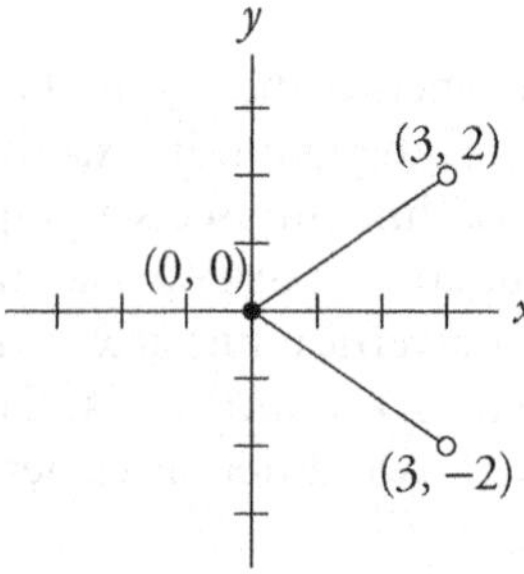

domain: $\{x | 0 \leq x < 3\}$ or $[0, 3)$
range: $\{y | -2 < y < 2\}$ or $(-2, 2)$
function: no

(e) $y = x^2$

domain: $\{x | x \in \mathbb{R}\}$ or $(-\infty, \infty)$

range: $\{y | y \geq 0\}$ or $[0, \infty)$

function: yes

(f) $x = y^2$

domain: $\{x | x \geq 0\}$ or $[0, \infty)$

range: $\{y | y \in \mathbb{R}\}$ or $(-\infty, \infty)$

function: no

Note that Examples 2(e) and (f) are illustrations of **inverse relations,** relations in which the domain and the range have been interchanged. Notice that while the relation in item (e) is a function, the inverse relation in item (f) is not.

Evaluating functions

This section will expand your knowledge of Common Core Mathematics functions represented as linear equations involving x and y, such as $y = x + 1$. A function can also be written using function notation, where $f(x)$ replaces y. This is read as "f of x." So, $f(x) = x + 1$, where the letter f represents the name of the function, x is a value from the function's domain, and $f(x)$ is a value from the function's range. Function notation is used to emphasize that the function value $f(x)$, or dependent variable, depends on the value of x, the independent variable.

The *values of a function* are the *values of the range* (or output) of the function. Given the function $f = \{(1, -3), (2, 4), (-1, 5), (3, -2)\}$, the value of the function at 1 is -3, at 2 is 4, and so forth. This is written $f(1) = -3$ and $f(2) = 4$ and is usually read, "f of $1 = -3$ and f of $2 = 4$." The lowercase letter f has been used here to indicate the concept of function, but any lowercase letter might have been used.

Example 3: Let $h = \{(3, 1), (2, 2), (1, -2), (-2, 3)\}$. Find each of the following.

(a) $h(3) =$	**(b)** $h(2) =$	**(c)** $h(1) =$	**(d)** $h(-2) =$
$h(3) = 1$	$h(2) = 2$	$h(1) = -2$	$h(-2) = 3$

When evaluating a function $f(x)$ for a specific value, substitute the x value into the equation to find the result.

Example 4: If $g(x) = 2x + 1$, find each of the following.

(a) $g(x) = 2x + 1$

If $x = -1$,

$$g(-1) = 2(-1) + 1$$
$$g(-1) = -2 + 1$$
$$g(-1) = -1$$

(b) $g(x) = 2x + 1$

If $x = 2$,

$$g(2) = 2(2) + 1$$
$$g(2) = 4 + 1$$
$$g(2) = 5$$

(c) $g(x) = 2x + 1$

If $x = a + 1$,

$$g(a+1) = 2(a+1) + 1$$
$$g(a+1) = 2a + 3$$

Example 5: If $f(x) = 3x^2 + x - 1$, find the range of f for the domain $\{-2, -1, 1\}$.

$$f(x) = 3x^2 + x - 1$$
$$f(-2) = 3(-2)^2 + (-2) - 1$$
$$f(-2) = 3(4) - 2 - 1$$
$$f(-2) = 12 - 3$$
$$f(-2) = 9$$

$$f(x) = 3x^2 + x - 1$$
$$f(-1) = 3(-1)^2 + (-1) - 1$$
$$f(-1) = 3(1) - 1 - 1$$
$$f(-1) = 3 - 2$$
$$f(-1) = 1$$

$$f(x) = 3x^2 + x - 1$$
$$f(1) = 3(1)^2 + (1) - 1$$
$$f(1) = 3(1) + 1 - 1$$
$$f(1) = 3$$

range: {9, 1, 3}

Example 6: A race car driver is traveling at an average speed of 90 miles per hour. The function $d(t) = 90t$ represents the distance the race car travels as a function of time in hours. If the race car driver travels for 16 minutes on the track, how many miles has he traveled?

Before evaluating the function, convert 16 minutes to hours.

$$\frac{16\ \cancel{\text{min}}}{1} \cdot \frac{1\ \text{hr}}{60\ \cancel{\text{min}}} = \frac{16}{60}\ \text{hr} = \frac{4}{15}\ \text{hr}$$

Substitute this value in for t below.

$$d(t) = 90t$$
$$d\left(\frac{4}{15}\right) = 90 \cdot \left(\frac{4}{15}\right)$$
$$d\left(\frac{4}{15}\right) = 24\ \text{miles}$$

The race car driver travels 24 miles in 16 minutes.

Variations

Common Core Standard: Interpreting Functions

Graph functions expressed symbolically and show key features of the graph, by hand in simple cases and using technology for more complicated cases (F.IF.7).

Common Core Standard: Creating Equations

Create equations in two or more variables to represent relationships between quantities; graph equations on coordinate axes with labels and scales (A.CED.2).

A **variation** is a relation that shows how one quantity changes relative to another quantity. A variation can be represented as an equation in which a constant relationship exists between the values of two variables.

Direct variation

In the equation $y = mx + b$, if m is a nonzero constant and $b = 0$, then you have the function $y = mx$ (often written $y = kx$), which is called a **direct variation.** That is, you can say that y varies directly with x or y is directly proportional to x. In this function, m (or k) is called the constant of proportionality or the constant of variation. The graph of every direct variation passes through the origin, and the values of the variables, x and y, either both increase at a constant rate or both decrease at a constant rate in most applications. An example of a direct variation is the relationship between the number of hours worked and the amount of money earned. The amount of money earned varies directly with the number of hours worked. Note that the graph of a direct variation is a linear function that has a slope of k and passes through the point (0, 0).

Example 7: Graph the direct variation equation $y = 2x$.

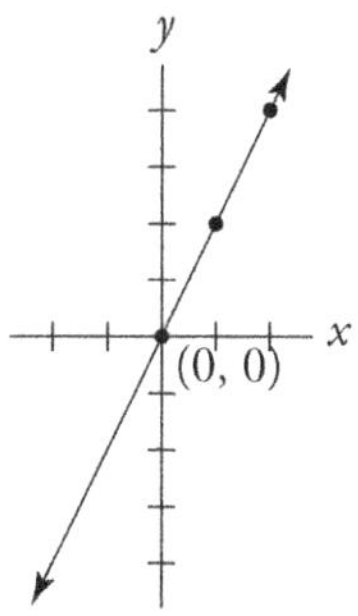

Example 8: If y varies directly with x, find the constant of variation when y is 2 and x is 4.

Because this is a direct variation,

$$y = kx \text{ (or } y = mx)$$

Now, replacing y with 2 and x with 4, we have

$$2 = k(4)$$

$$k = \frac{2}{4} \text{ or } k = \frac{1}{2}$$

The constant of variation is $\frac{1}{2}$.

Example 9: If y varies directly with x and the constant of variation is 2, find y when x is 6.

Since this is a direct variation, simply replace k with 2 and x with 6 in the following equation.

$$y = kx$$

$$y = 2(6)$$

$$y = 12$$

A direct variation can also be written as a proportion.

$$\frac{y_1}{x_1} = \frac{y_2}{x_2}$$

This proportion is read, "y_1 is to x_1 as y_2 is to x_2." x_1 and y_2 are called the **means,** and y_1 and x_2 are called the **extremes.** Using the cross-products property, the product of the means is always equal to the product of the extremes. You can solve a proportion by simply multiplying the means and extremes and then solving as usual.

Example 10: Suppose r varies directly with p. If r is 3 when p is 7, find p when r is 9.

Method 1. Using proportions:

Set up the direct variation proportion.

$$\frac{r_1}{p_1} = \frac{r_2}{p_2}$$

Now, substitute in the values.

$$\frac{3}{7} = \frac{9}{p}$$

Multiplying the means and extremes gives

$$3p = 63$$
$$p = 21$$

Method 2. Using $y = kx$:

Substitute the y with p and the x with r.

$$p = kr$$

Use the first set of information and substitute 3 for r and 7 for p, then find k.

$$7 = k(3)$$
$$\frac{7}{3} = k \text{ or } k = \frac{7}{3}$$

Rewrite the direct variation equation as $p = \frac{7}{3}r$.

Now use the second set of information that says r is 9, substitute this into the preceding equation, and solve for p.

$$p = \frac{7}{3}r$$
$$p = \frac{7}{\cancel{3}_1}\left(\frac{\cancel{9}^3}{1}\right) = 21$$

Inverse variation (indirect variation)

A variation where $y = \frac{m}{x}$ or $y = \frac{k}{x}$ is called an **inverse variation** (or *indirect variation*). That is, in most applications, as x *increases,* y *decreases.* And as y *increases,* x *decreases.* You may see the equation $xy = k$ representing an inverse variation, but this is simply a rearrangement of $y = \frac{k}{x}$. Note that the constant of variation, $k \neq 0$, and the ordered pair (x, y) satisfy the inverse variation. This function is also referred to as an *inverse* or *indirect proportion.* Again, m (or k) is called the constant of variation. An example of an indirect variation is the relationship between the number of hours it takes to build a house and the number of people building the house. The amount of time varies indirectly with the number of people building the house because as the number of workers increases, the length of time to complete the house will decrease. Note that the graph of an inverse variation is a rational function.

Example 11: If y varies indirectly with x, find the constant of variation when y is 2 and x is 4.

Since this is an indirect or inverse variation,

$$y = \frac{k}{x}$$

Now, substituting y with 2 and x with 4,

$$2 = \frac{k}{4}$$

$$k = 2(4) \text{ or } 8$$

The constant of variation is 8.

Example 12: If y varies indirectly with x and the constant of variation is 2, find y when x is 6.

Since this is an indirect variation, simply substitute k with 2 and x with 6 in the following equation.

$$y = \frac{k}{x}$$

$$y = \frac{2}{6}$$

$$y = \frac{1}{3}$$

As with direct variation, inverse variation also can be written as a proportion.

$$\frac{y_1}{x_2} = \frac{y_2}{x_1}$$

Notice that in the inverse proportion, the x_1 and the x_2 switched their positions from the direct variation proportion.

Example 13: If y varies indirectly with x and $y = 4$ when $x = 9$, find x when $y = 3$.

Method 1. Using proportions:

Set up the indirect variation proportion.

$$\frac{y_1}{x_2} = \frac{y_2}{x_1}$$

Now, substitute in the values.

$$\frac{4}{x} = \frac{3}{9}$$

Multiplying the means and extremes gives

$$3x = 36$$
$$x = 12$$

Method 2. Using $y = \frac{k}{x}$:

Use the first set of information and substitute 4 for y and 9 for x, then find k.

$$4 = \frac{k}{9}$$
$$36 = k \text{ or } k = 36$$

Rewrite the inverse variation equation as $xy = 36$ or $y = \frac{36}{x}$.

Now use the second set of information that says y is 3, substitute this into the preceding equation, and solve for x.

$$3 = \frac{36}{x}$$
$$3x = 36$$
$$x = 12$$

Chapter Check-Out

Questions

1. True or false: The graph of a direct variation sometimes passes through the origin.

Question 2 refers to the following set.

$$\{(-2, 1), (-1, 3), (2, 4), (3, 5)\}$$

2. (a) Identify the domain of the relation.
   (b) Identify the range of the relation.
   (c) Answer yes or no to determine whether the relation is a function.

Questions 3–5 refer to the following information.

On a cold winter day in Buffalo, New York, it starts to snow at a rate of 1 inch every 2 hours. It snowed for 6 hours, but there was already 4 inches of snow on the ground when the snow started.

3. Draw a graph showing the depth of the snow $D(t)$ in inches as a function of the time (t) in hours.
4. Find and interpret the domain and the range.
5. Is the function a linear function? Explain your answer.
6. The function $C(x) = 2.50x + 3$ represents the cost $C(x)$ in dollars of a taxi ride of x miles. If you only have $20.50 to pay for the ride, how many miles can you ride in the taxi from point A to point B?
7. Explain why the graph of $x = 5$ is not a function.
8. Which of the following three figures are graphs of functions?

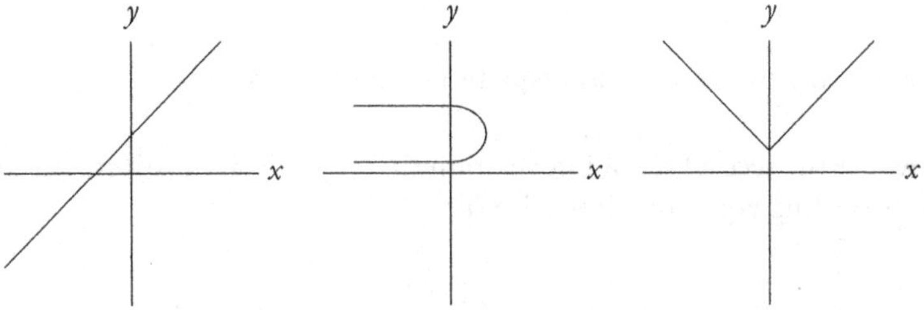

(a) (b) (c)

9. If $g(x) = 2x^2 + x + 1$, what is $g(-2)$?
10. If the domain of $f(x) = x^2 + 2x + 2$ is $\{1, 2, 3\}$, what is the range?

Questions 11–13 refer to the following information.

Evaluate each of the following equations, and determine if each equation represents a direct variation, inverse variation, or neither. If it is a direct variation or an inverse variation, find the constant of variation.

11. $xy = 5$
12. $3y = 6 + 2x$
13. $3y - 5x = 0$
14. If y varies directly with x, find the constant of variation if y is 4 when x is 12. Then write the equation for the direct variation.
15. r varies directly with p. If r is 6 and p is 11, find p when r is 18.
16. If y varies inversely with x and the constant of variation is 3, find y when x is 7. Then write an equation for the inverse variation.
17. If y varies inversely with x and y is 18 when x is 4, find y when x is 12.

Questions 18 and 19 refer to the following information.

The perimeter, P, of a square varies directly with the length of each side, s, of the square.

18. Write an equation that relates P and s.
19. What is the constant of variation?
20. Suppose it takes a group of 5 people 2 days to rake leaves in a park. How long would it take 8 people to rake leaves in the same park?

Answers

1. False; the graph of a direct variation always passes through the origin.
2. **(a)** domain: $\{-2, -1, 2, 3\}$

   (b) range: $\{1, 3, 4, 5\}$

   (c) function: Yes. The relation represents a function since each value of the domain is paired with exactly one value of the range.

3.

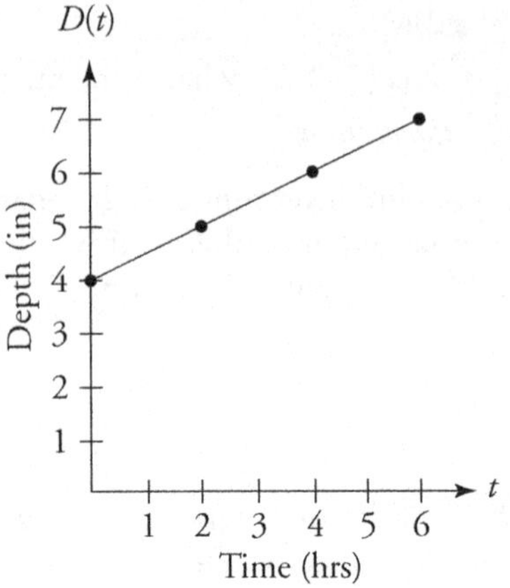

4. The domain is $0 \le x \le 6$, which represents the time (in hours) it snows. The range is $4 \le D(t) \le 7$, which represents the amount of snow (in inches) on the ground.

5. Yes, the function is linear since there is a constant rate of change.

6. 7 miles

7. The graph $x = 5$ is not a function because it does not pass the vertical line test.

8. Graphs (a) and (c)

9. 7

10. $\{5, 10, 17\}$

11. inverse variation; $k = 5$

12. neither

13. direct variation; $k = \frac{5}{3}$

14. $k = \frac{1}{3}$; $y = \frac{1}{3}x$

15. 33

16. $\frac{3}{7}$; $y = \frac{3}{x}$

17. 6

18. $P = 4s$

19. The constant of variation is 4.

20. 1.25 days

Chapter 10

SIMPLIFYING RADICALS

Chapter Check-In

- ❑ Simplifying square roots
- ❑ Operations with square roots

Common Core Mathematics explains exponential relationships with expressions of radicals and rational exponents to represent solutions to algebraic equations. Simplifying radicals was first introduced in Chapter 1, "Foundations for Understanding Algebra."

The symbol $\sqrt{\ }$ is called a **radical sign** and is used to designate **square root.** A *radical expression* is an expression with a radical sign, where the expression underneath the radical is the *radicand.* To designate **cube root,** a small three (the index) is placed above the radical sign, $\sqrt[3]{\ }$. Radicals with higher-index roots can be written in a similar manner. When two radical signs are next to each other, it automatically means that the two are multiplied; the multiplication sign may be omitted. Note that the square root of a negative number is not possible within the real number system; a completely different system of **imaginary numbers** is used. The (so-called) imaginary numbers are multiples of the imaginary unit *i*:

$$\sqrt{-1} = i,\ \sqrt{-4} = 2i,\ \sqrt{-9} = 3i,\ \text{and so on}$$

Simplifying Square Roots

Common Core Standard: The Real Number System

Explain why the sum or product of two rational numbers is rational; that the sum of a rational number and an irrational number is an irrational; and that the product of a nonzero rational number and an irrational number is irrational (N.RN.3).

Common Core Standard: Reasoning with Equations and Inequalities

Solve simple rational and radical equations in one variable, and give examples showing how extraneous solutions may arise (A.REI.2).

A simplified radical expression is one in which the radicand contains no perfect square factors other than 1, the denominator of a fraction contains no radicals, and there are no fractions in the radicand. Simplifying non-perfect square roots involves removing perfect square factors and is explained in more detail in Chapter 1. To simplify square roots that contain variables, look for variables with even exponents, which are perfect squares.

For example, if a is a nonnegative number, a^4 is a perfect square because $a^2 \cdot a^2 = a^4$. So, we can conclude that $\sqrt{a^4} = a^2$. If the variable does not have an even exponent, write an equivalent expression that contains the product of a perfect square and the variable in the radicand. For example, a^3 can be written as $a^2 \cdot a$, so $\sqrt{a^3} = \sqrt{a^2(a)} = \sqrt{a^2} \cdot \sqrt{a} = a\sqrt{a}$.

Example 1: Simplify.

(a) $\sqrt{9} = 3$

(b) $-\sqrt{9} = -3$

(a) and (b): Reminder: This notation is used in many texts and will be adhered to in this book.

(c) $\sqrt{18} = \sqrt{9 \cdot 2} = \sqrt{9} \cdot \sqrt{2} = 3\sqrt{2}$

When simplifying radical expressions that contain variables, assume each variable is nonnegative. A variable that contains an even exponent is a perfect square. A variable that contains an odd exponent can be written as the product of a perfect square and the variable itself. For example, $\sqrt{x^5} = \sqrt{x^4 \cdot x} = \sqrt{x^4} \cdot \sqrt{x} = x^2\sqrt{x}$.

Example 2: Simplify (each variable is nonnegative).

(a) $\sqrt{x^4} = x^2$

(b) $\sqrt{x^6 y^8} = \sqrt{x^6}\sqrt{y^8} = x^3 y^4$

(c) $\sqrt{25a^4b^6} = \sqrt{25}\sqrt{a^4}\sqrt{b^6} = 5a^2b^3$

(d) $\sqrt{x^7} = \sqrt{x^6(x)} = \sqrt{x^6}\sqrt{x} = x^3\sqrt{x}$

(e) $\sqrt{x^9y^8} = \sqrt{x^9}\sqrt{y^8} = \sqrt{x^8(x)}\sqrt{y^8} = \sqrt{x^8}\sqrt{x}\sqrt{y^8} = x^4\sqrt{x}\cdot y^4 = x^4y^4\sqrt{x}$

(f) $\sqrt{16x^5} = \sqrt{16}\sqrt{x^5} = \sqrt{16}\sqrt{x^4(x)} = \sqrt{16}\sqrt{x^4}\sqrt{x} = 4x^2\sqrt{x}$

Operations with Square Roots

Common Core Standard: The Real Number System

Explain why the sum or product of two rational numbers is rational; that the sum of a rational number and an irrational number is irrational; and that the product of a nonzero rational number and irrational number is irrational (N.RN.3).

Common Core Standard: Reasoning with Equations and Inequalities

Solve simple rational and radical equations in one variable, and give examples showing how extraneous solutions may arise (A.REI.2).

You can use properties of real numbers to perform a number of different operations with square roots. Some of these operations involve a single radical sign, while others can involve multiple radical signs. The rules governing these operations should be carefully reviewed.

Under a single radical sign

You may perform operations *under a single radical sign.*

Example 3: Perform the operation indicated.

(a) $\sqrt{(5)(20)} = \sqrt{100} = 10$

(b) $\sqrt{30+6} = \sqrt{36} = 6$

(c) $\sqrt{\frac{32}{2}} = \sqrt{16} = 4 \quad \left(\text{Note: } \sqrt{\frac{32}{2}} = \frac{\sqrt{32}}{\sqrt{2}}\right)$

(d) $\sqrt{30-5} = \sqrt{25} = 5$

(e) $\sqrt{2+5} = \sqrt{7}$

When radical values are alike

You can find the sum or difference of two or more radical expressions only if the expression contains like radicals. Like radicals contain the same radicand, such as $2\sqrt{3}$ and $-7\sqrt{3}$. The process of combining like radicals to simplify sums or differences involves the use of the distributive property, where you add or subtract the coefficients (numbers in front of the radical sign) and keep the original radicand.

Example 4: Perform the operation indicated.

(a) $2\sqrt{3}+3\sqrt{3}=(2+3)\sqrt{3}=5\sqrt{3}$

(b) $4\sqrt{6}-2\sqrt{6}=(4-2)\sqrt{6}=2\sqrt{6}$

(c) $5\sqrt{2}+\sqrt{2}=5\sqrt{2}+1\sqrt{2}=(5+1)\sqrt{2}=6\sqrt{2}$

Note that the coefficient 1 is understood in $\sqrt{2}$.

When radical values are different

Unlike radicals contain different radicands, such as $2\sqrt{5}$ and $-7\sqrt{3}$, and cannot be combined or simplified.

Example 5:

(a) $\sqrt{28}-\sqrt{3}\neq\sqrt{25}$

(b) $\sqrt{16}+\sqrt{9}\neq\sqrt{25}$

$$4+3\neq 5$$

(c) $3\sqrt{7}+5\sqrt{3}\neq 8\sqrt{10}$

Addition and subtraction of square roots after simplifying

Sometimes, after simplifying the radical expression, it is possible to further simplify using addition or subtraction. Always simplify further if possible.

Example 6: Simplify the expression.

(a) $\sqrt{50}+3\sqrt{2}=$

These cannot be added until $\sqrt{50}$ is simplified.

$$\sqrt{50}=\sqrt{25\cdot 2}=\sqrt{25}\cdot\sqrt{2}=5\sqrt{2}$$

Now that there are like radicals,

$$5\sqrt{2}+3\sqrt{2}=(5+3)\sqrt{2}=8\sqrt{2}$$

(b) $\sqrt{300}+\sqrt{12}=$

Try to simplify each one.

$$\sqrt{300}=\sqrt{100\cdot 3}=\sqrt{100}\cdot\sqrt{3}=10\sqrt{3}$$
$$\sqrt{12}=\sqrt{4\cdot 3}=\sqrt{4}\cdot\sqrt{3}=2\sqrt{3}$$

Now that there are like radicals,

$$10\sqrt{3}+2\sqrt{3}=(10+2)\sqrt{3}=12\sqrt{3}$$

Products of nonnegative square roots

To simplify radical expressions involving products, refer to the multiplication property of square roots, which states that if $x \geq 0$ and $y \geq 0$, then $\sqrt{xy}=\sqrt{x}\cdot\sqrt{y}$. Remember that in multiplication of radicals, the multiplication sign may be omitted. Always simplify the answer when possible.

Example 7: Simplify each radical expression.

(a) $\sqrt{2}\cdot\sqrt{8}=\sqrt{16}=4$

(b) If each variable is nonnegative,

$$\sqrt{x^3}\cdot\sqrt{x^5}=\sqrt{x^8}=x^4$$

(c) If each variable is nonnegative,

$$\sqrt{ab}\cdot\sqrt{ab^3c}=\sqrt{a^2b^4c}=\sqrt{a^2}\sqrt{b^4}\sqrt{c}=ab^2\sqrt{c}$$

(d) If each variable is nonnegative,

$$\begin{aligned}\sqrt{3x}\cdot\sqrt{6xy^2}\cdot\sqrt{2xy}&=\sqrt{36x^3y^3}\\&=\sqrt{36}\sqrt{x^3}\sqrt{y^3}\\&=\sqrt{36}\sqrt{x^2(x)}\sqrt{y^2(y)}\\&=\sqrt{36}\sqrt{x^2}\sqrt{x}\sqrt{y^2}\sqrt{y}=6xy\sqrt{xy}\end{aligned}$$

(e) $2\sqrt{5}\cdot 7\sqrt{3}=(2\cdot 7)\sqrt{5\cdot 3}=14\sqrt{15}$

Quotients of nonnegative square roots

To simplify radical expressions involving quotients, refer to the division property of square roots, which states that if $x \geq 0$ and $y > 0$, then $\frac{\sqrt{x}}{\sqrt{y}} = \sqrt{\frac{x}{y}}$.

Example 8: Simplify each radical expression. All variables are assumed to be nonnegative.

(a) $\frac{\sqrt{10}}{\sqrt{2}} = \sqrt{\frac{10}{2}} = \sqrt{5}$

(b) $\frac{\sqrt{24}}{\sqrt{3}} = \sqrt{\frac{24}{3}} = \sqrt{8} = 2\sqrt{2}$

(c) $\frac{\sqrt{28x^6}}{\sqrt{7x^2}} = \sqrt{\frac{28x^6}{7x^2}} = \sqrt{4x^4} = 2x^2$

(d) $\frac{\sqrt{15}}{\sqrt{6}} = \sqrt{\frac{15}{6}} = \sqrt{\frac{5}{2}}$ or $\frac{\sqrt{5}}{\sqrt{2}}$

Note that the radical expression in item (d) is not in simplified form since there is a radical in the denominator. To fully simplify the expression, rationalize the denominator, which removes the radical in the denominator. In order to rationalize the denominator, multiply the numerator and denominator by a radical that will make the radicand in the denominator a perfect square. Multiply by 1 in the form of $1 = \frac{\sqrt{2}}{\sqrt{2}}$.

$$\frac{\sqrt{5}}{\sqrt{2}} \cdot 1 = \frac{\sqrt{5}}{\sqrt{2}} \cdot \frac{\sqrt{2}}{\sqrt{2}} = \frac{\sqrt{10}}{\sqrt{4}} = \frac{\sqrt{10}}{2}$$

Example 9: Simplify each expression. All variables are assumed to be nonnegative.

(a) $\frac{5\sqrt{7}}{\sqrt{12}}$

First simplify $\sqrt{12}$:

$$\frac{5\sqrt{7}}{\sqrt{12}} = \frac{5\sqrt{7}}{2\sqrt{3}} \cdot 1 = \frac{5\sqrt{7}}{2\sqrt{3}} \cdot \frac{\sqrt{3}}{\sqrt{3}} = \frac{5\sqrt{21}}{2\sqrt{9}} = \frac{5\sqrt{21}}{2 \cdot 3} = \frac{5\sqrt{21}}{6}$$

or

$$\frac{5\sqrt{7}}{\sqrt{12}}\cdot\frac{\sqrt{12}}{\sqrt{12}}=\frac{5\sqrt{7}\cdot\sqrt{12}}{12}\cdot=\frac{5\sqrt{84}}{12}=\frac{5\sqrt{4\cdot 21}}{12}=\frac{10\sqrt{21}}{12}=\frac{5\sqrt{21}}{6}$$

(b) $\frac{9\sqrt{2x}}{\sqrt{24x^3}}=9\sqrt{\frac{2x}{24x^3}}=9\sqrt{\frac{1}{12x^2}}=\frac{9}{\sqrt{12x^2}}=\frac{9}{2x\sqrt{3}}\cdot 1=\frac{9}{2x\sqrt{3}}\cdot\frac{\sqrt{3}}{\sqrt{3}}=\frac{9\sqrt{3}}{2x\cdot 3}$

$$=\frac{9\sqrt{3}}{6x}=\frac{3\sqrt{3}}{2x}$$

In order to leave a rational term in the denominator, it is sometimes necessary to multiply both the numerator and denominator by the **conjugate** of the denominator. The conjugate of a binomial contains the same terms but the opposite sign. Thus, $(x + y)$ and $(x - y)$ are conjugates. Note that the product of conjugates is the difference of squares.

Example 10: Simplify each expression.

(a) $\frac{3}{2+\sqrt{3}}\cdot 1=\frac{3}{2+\sqrt{3}}\cdot\frac{\left(2-\sqrt{3}\right)}{\left(2-\sqrt{3}\right)}=\frac{3\left(2-\sqrt{3}\right)}{4-3}=\frac{6-3\sqrt{3}}{1}=6-3\sqrt{3}$

(b) $\frac{1+\sqrt{5}}{2-\sqrt{5}}\cdot 1=\frac{\left(1+\sqrt{5}\right)}{\left(2-\sqrt{5}\right)}\cdot\frac{\left(2+\sqrt{5}\right)}{\left(2+\sqrt{5}\right)}=\frac{2+\sqrt{5}+2\sqrt{5}+\sqrt{25}}{4+2\sqrt{5}-2\sqrt{5}-\sqrt{25}}=\frac{2+3\sqrt{5}+5}{4-5}$

$$=\frac{7+3\sqrt{5}}{-1}=-7-3\sqrt{5}$$

Chapter Check-Out

Questions

1. Describe and correct the error in simplifying the following expression.

$$\sqrt{80}=\sqrt{4\cdot 20}=\sqrt{4}\cdot\sqrt{20}=2\sqrt{20}$$

2. If each variable is nonnegative, then $\sqrt{16a^6b^8}=$ _____.

3. If each variable non negative, then $\sqrt{x^7}=$ _____.

For question 4, select the best answer from among the four choices.

4. Which expression is in its simplest form?

A. $4\sqrt{8}$

B. $\dfrac{\sqrt{15}}{2}$

C. $\dfrac{5}{\sqrt{6}}$

D. $\sqrt{\dfrac{1}{4}}$

For question 5, select the best answer from among the four choices.

5. Which expression is equivalent to $\sqrt{60}+2\sqrt{15}$?

A. $6\sqrt{15}$

B. $4\sqrt{15}$

C. $4\sqrt{30}$

D. $3\sqrt{75}$

For questions 6–8, write each expression in simplest form.

6. $\sqrt{(4)(36)} =$

7. $\sqrt{6}\times\sqrt{10} =$

8. $\sqrt{5}\left(\sqrt{8}-3\sqrt{2}\right)$

9. If x and y are nonnegative, then $\sqrt{x^3 y}\cdot\sqrt{x^2 y} =$

Question 10 refers to the figure below.

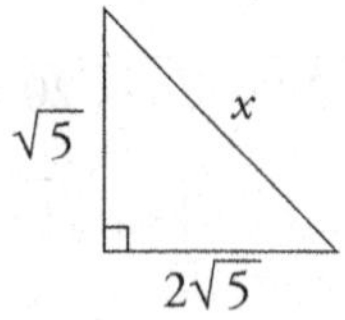

10. Find the missing length, x, in the triangle above.

For questions 11–13, express answers in simplest form with rationalized denominators.

11. $\frac{\sqrt{28}}{\sqrt{7}} =$

12. $\frac{3\sqrt{5}}{\sqrt{2}} =$

13. $\frac{2}{3+\sqrt{2}} =$

Answers

1. Since the factors 4 and 20 do not contain the largest perfect square, $\sqrt{20}$ is not in simplest form. The factors 5 and 16 contain the largest perfect square of 80, so $\sqrt{80} = \sqrt{16 \cdot 5} = \sqrt{16} \cdot \sqrt{5} = 4\sqrt{5}$.

2. $4a^3b^4$

3. $x^3\sqrt{x}$

4. B

5. B

6. 12

7. $2\sqrt{15}$

8. $-1\sqrt{10}$

9. $x^2y\sqrt{x}$

10. $x = 5$

11. 2

12. $\frac{3\sqrt{10}}{2}$

13. $\frac{6-2\sqrt{2}}{7}$

Chapter 11
QUADRATIC EQUATIONS

Chapter Check-In

- ❑ Solving quadratic equations
- ❑ The factoring method
- ❑ The quadratic formula
- ❑ Completing the square

Common Core Mathematics asks you to understand and solve quadratic equations as a process of reasoning. A **quadratic equation** is an equation that can be written in standard form as $ax^2 + bx + c = 0$, where a, b, and c are real numbers and $a \neq 0$.

There are many methods to solve quadratics. These methods include graphing to find the zeros (x-intercepts) of the function, or using the algebraic methods described in this chapter. The solution to a quadratic equation is any number(s) that will make the equation *true*. There could be two solutions, one solution, or no real number solutions. The solution(s) of a quadratic equation are also called the *roots of the equation* or *zeros of the function.*

Solving Quadratic Equations

Common Core Standard: Arithmetic with Polynomials and Rational Expressions

Identify zeros of polynomials when suitable factorizations are available, and use the zeros to construct a rough graph of the function defined by the polynomial (A.APR.3).

Common Core Standard: Creating Equations

Create equations and inequalities in one variable and use them to solve problems. Include equations arising from linear and quadratic functions, and simple rational and exponential functions (A.CED.1). Represent constraints by equations or inequalities and by systems of equations and/or inequalities, and interpret solutions as viable or non-viable options in a modeling context (A.CED.3). Rearrange formulas to highlight a quantity of interest using the same reasoning as in solving equations (A.CED.4).

Common Core Standard: Reasoning with Equations and Inequalities

Solve quadratic equations in one variable (A.REI.4). Use the method of completing the square to transform any quadratic equation in x into an equation of the form $(x - p)^2 = q$ that has the same solutions. Derive the quadratic formula from this form (A.REI.4.a). Solve quadratic equations by inspection, taking square roots, completing the square, the quadratic formula, and factoring, as appropriate to the initial form of the equation. Recognize when the quadratic formula gives complex solutions and write them as $a \pm bi$ for real numbers a and b (A.REI.4.b).

Common Core Standard: Seeing Structure in Expressions

Interpret parts of an expression, such as terms, factors, and coefficients (A.SSE.1.a). Choose and produce an equivalent form of an expression to reveal and explain properties of the quantity represented by the expression (A.SSE.3). Factor a quadratic expression to reveal the maximum or minimum value of the function it defines (A.SSE.3.a).

There are three basic methods for solving quadratic equations algebraically: factoring, using the quadratic formula, and completing the square. For a review of factoring, refer to Chapter 5, "Monomials, Polynomials, and Factoring."

Factoring

To solve a quadratic equation by factoring, follow these steps:

1. Write the quadratic equation in standard form.
2. Factor.
3. Split the product of the factors into separate equations by setting each factor equal to zero. This is called the **zero-product property.**

4. Solve each of these equations.
5. Check by substituting your solution(s) in the original quadratic equation.

Example 1: Solve for x: $x^2 - 6x = 16$.

Following the steps,

$$x^2 - 6x = 16 \text{ becomes } x^2 - 6x - 16 = 0$$

Factor.

$$(x - 8)(x + 2) = 0$$

Using the zero-product property,

$$\begin{aligned} x - 8 &= 0 \quad \text{or} \quad & x + 2 &= 0 \\ x &= 8 & x &= -2 \end{aligned}$$

Then to check,

$$\begin{aligned} 8^2 - 6(8) &= 16 \quad \text{or} \quad & (-2)^2 - 6(-2) &= 16 \\ 64 - 48 &= 16 & 4 + 12 &= 16 \\ 16 &= 16✓ & 16 &= 16✓ \end{aligned}$$

Both values, 8 and −2, are solutions to the original equation.

Example 2: Solve for y: $y^2 = -6y - 5$.

Writing the quadratic equation in standard form,

$$y^2 + 6y + 5 = 0$$

Factor.

$$(y + 5)(y + 1) = 0$$

Using the zero-product property,

$$\begin{aligned} y + 5 &= 0 \quad \text{or} \quad & y + 1 &= 0 \\ y &= -5 & y &= -1 \end{aligned}$$

To check, $y^2 = -6y - 5$,

$$(-5)^2 = -6(-5) - 5 \quad \text{or} \quad (-1)^2 = -6(-1) - 5$$
$$25 = 30 - 5 \qquad\qquad 1 = 6 - 5$$
$$25 = 25✓ \qquad\qquad 1 = 1✓$$

A quadratic with a term missing is called an **incomplete quadratic equation** (as long as the ax^2 term isn't missing).

Example 3: Solve for x: $x^2 - 16 = 0$.

Factor.

$$(x + 4)(x - 4) = 0$$

$$x + 4 = 0 \quad \text{or} \quad x - 4 = 0$$
$$x = -4 \qquad\qquad x = 4$$

To check,

$$x^2 - 16 = 0 \qquad\qquad x^2 - 16 = 0$$
$$(-4)^2 - 16 = 0 \quad \text{or} \quad (4)^2 - 16 = 0$$
$$16 - 16 = 0 \qquad\qquad 16 - 16 = 0$$
$$0 = 0✓ \qquad\qquad 0 = 0✓$$

When the quadratic formula is in the form $ax^2 + c = 0$, where a and c are real numbers and $a \neq 0$, you can also solve the problem by isolating the variable term on one side of the equation and simplify by using square roots.

$$x^2 - 16 = 0$$
$$x^2 = 16$$
$$\sqrt{x^2} = \sqrt{16}$$
$$x = 4 \text{ or } x = -4$$

Example 4: Solve for x: $x^2 + 6x = 0$.

Factor.

$$x(x + 6) = 0$$

$$x = 0 \quad \text{or} \quad x + 6 = 0$$
$$x = 0 \qquad\qquad x = -6$$

To check,

$$\begin{aligned} x^2+6x&=0 \\ (0)^2+6(0)&=0 \\ 0+0&=0 \\ 0&=0✓ \end{aligned} \quad \text{or} \quad \begin{aligned} x^2+6x&=0 \\ (-6)^2+6(-6)&=0 \\ 36+(-36)&=0 \\ 0&=0✓ \end{aligned}$$

Example 5: Solve for x: $2x^2 + 2x - 1 = x^2 + 6x - 5$.

First, simplify by moving all terms to one side of the equation and combining like terms.

$$\begin{array}{rr} 2x^2+2x-1= & x^2+6x-5 \\ \underline{-x^2-6x+5} & \underline{-x^2-6x+5} \\ x^2-4x+4= & 0 \end{array}$$

Now, factor.

$$\begin{aligned} (x-2)(x-2)&=0 \\ x-2&=0 \\ x&=2 \end{aligned}$$

To check,

$$\begin{aligned} 2x^2+2x-1&=x^2+6x-5 \\ 2(2)^2+2(2)-1&=(2)^2+6(2)-5 \\ 8+4-1&=4+12-5 \\ 11&=11✓ \end{aligned}$$

The quadratic formula

Many quadratic equations cannot be solved by factoring. This is generally true when the roots, or solution(s), are not rational numbers. A second method of solving quadratic equations involves the use of the **quadratic formula:**

$$x=\frac{-b\pm\sqrt{b^2-4ac}}{2a}$$

a, b, and c are taken from the quadratic equation written in its general form of $ax^2 + bx + c = 0$, where a is the coefficient of x^2, b is the coefficient of x, and c is the constant. It is important to note that the quadratic formula can be used to solve any quadratic equation, especially when a quadratic equation can't be factored easily.

When using the quadratic formula, you should be aware of three possibilities. These three possibilities are distinguished by a part of the formula called the discriminant. The **discriminant** is the value under the radical sign, $b^2 - 4ac$. A quadratic equation with real numbers as coefficients can have the following:

- Two different real roots if the discriminant $b^2 - 4ac$ is a positive number
- One real root if the discriminant $b^2 - 4ac$ is equal to 0
- No real roots if the discriminant $b^2 - 4ac$ is a negative number

Example 6: Solve for x: $x^2 - 5x = -6$.

Setting all terms equal to 0, you have

$$x^2 - 5x + 6 = 0$$

Then substitute 1 (which is understood to be in front of the x^2), -5, and 6 for a, b, and c, respectively, in the quadratic formula and simplify.

$$x = \frac{-b \pm \sqrt{b^2 - 4ac}}{2a}$$

$$x = \frac{-(-5) \pm \sqrt{(-5)^2 - 4(1)(6)}}{2(1)}$$

$$x = \frac{5 \pm \sqrt{25 - 24}}{2}$$

$$x = \frac{5 \pm \sqrt{1}}{2}$$

$$x = \frac{5 \pm 1}{2}$$

$$x = \frac{5+1}{2} \text{ or } x = \frac{5-1}{2}$$

$$x = \frac{6}{2} \text{ or } x = \frac{4}{2}$$

$$x = 3 \text{ or } x = 2$$

Because the discriminant $b^2 - 4ac$ is positive, there are two different real roots.

Example 6 produces rational roots. In Example 7, the quadratic formula is used to solve an equation whose roots are not rational.

Example 7: Solve for y: $y^2 = -2y + 2$.

Setting all terms equal to 0, you have

$$y^2 + 2y - 2 = 0$$

Then substitute 1, 2, and −2 for a, b, and c, respectively, in the quadratic formula and simplify.

$$y = \frac{-b \pm \sqrt{b^2 - 4ac}}{2a}$$

$$y = \frac{-(2) \pm \sqrt{(2)^2 - 4(1)(-2)}}{2(1)}$$

$$y = \frac{-2 \pm \sqrt{4+8}}{2}$$

$$y = \frac{-2 \pm \sqrt{12}}{2}$$

$$y = \frac{-2 \pm \sqrt{4}\sqrt{3}}{2}$$

$$y = \frac{-2 \pm 2\sqrt{3}}{2}$$

$$y = \frac{\overset{1}{\cancel{2}}\left(-1 \pm \sqrt{3}\right)}{\underset{1}{\cancel{2}}}$$

$$y = -1 + \sqrt{3} \text{ or } y = -1 - \sqrt{3}$$

Note that the two roots are irrational.

Example 8: Solve for x: $x^2 + 2x + 1 = 0$.

Substituting $a = 1$, $b = 2$, and $c = 1$ into the quadratic formula,

$$x = \frac{-b \pm \sqrt{b^2 - 4ac}}{2a}$$

$$x = \frac{-(2) \pm \sqrt{(2)^2 - 4(1)(1)}}{2(1)}$$

$$x = \frac{-2 \pm \sqrt{4-4}}{2}$$

$$x = \frac{-2 \pm \sqrt{0}}{2}$$

$$x = \frac{-2}{2} = -1$$

Since the discriminant $b^2 - 4ac$ is 0, the equation has one root.

The quadratic formula can also be used to solve quadratic equations whose roots are imaginary numbers; that is, they have no solution in the real number system.

Example 9: Solve for x: $x(x + 2) + 2 = 0$, or $x^2 + 2x + 2 = 0$.

Substituting $a = 1$, $b = 2$, and $c = 2$ into the quadratic formula,

$$x = \frac{-b \pm \sqrt{b^2 - 4ac}}{2a}$$

$$x = \frac{-(2) \pm \sqrt{(2)^2 - 4(1)(2)}}{2(1)}$$

$$x = \frac{-2 \pm \sqrt{4-8}}{2}$$

$$x = \frac{-2 \pm \sqrt{-4}}{2}$$

Since the discriminant $b^2 - 4ac$ is negative, this equation has no solution in the real number system.

But if you were to express the solution using imaginary numbers, as discussed at the beginning of Chapter 10, "Simplifying Radicals," the solutions would be $x=\frac{-2\pm\sqrt{-4}}{2}=\frac{-2\pm 2i}{2}=-1\pm i$ or $-1+i,\ -1-i$.

Completing the square

A third method of solving quadratic equations that works with both real and imaginary roots is called completing the square. This method is best to use when the coefficient of x^2 is 1 and the quadratic equation does not factor easily. To solve a quadratic equation by completing the square, follow these steps:

1. Write the equation in the form $ax^2 + bx = -c$.
2. Make sure that $a = 1$ (if $a \neq 1$, multiply each term in the equation by $\frac{1}{a}$ before proceeding).
3. Using the value of b from this new equation, add $\left(\frac{b}{2}\right)^2$ to both sides of the equation to form a perfect square trinomial on the left side of the equation. Factor the perfect square trinomial.
4. Find the square root of both sides of the equation.
5. Solve the resulting equation.

Example 10: Solve for x: $x^2 - 6x + 5 = 0$.

Arrange in the form of $ax^2 + bx = -c$:

$$x^2 - 6x = -5$$

Because $a = 1$, add $\left(\frac{-6}{2}\right)^2$, or 9, to both sides to complete the square.

$$x^2-6x+9=-5+9$$

$$x^2-6x+9=4$$

$$(x-3)^2=4$$

Take the square root of both sides.

$$x - 3 = \pm 2$$

Solve.

$$x-3=\pm 2$$

$$\begin{array}{rcr} x-3=+2 & \text{or} & x-3=-2 \\ +3\quad +3 & & +3\quad +3 \\ \hline x=5 & \text{or} & x=1 \end{array}$$

Example 11: Solve for y: $y^2 + 2y - 4 = 0$.

Arrange in the form of $ay^2 + by = -c$:

$$y^2 + 2y = 4$$

Because $a = 1$, add $\left(\frac{2}{2}\right)^2$, or 1, to both sides to complete the square.

$$y^2+2y+1=4+1$$
$$y^2+2y+1=5$$
$$(y+1)^2=5$$

Take the square root of both sides.

$$y+1=\pm\sqrt{5}$$

Solve.

$$\begin{array}{l} y+1=\quad \pm\sqrt{5} \\ \;\;-1\quad -1 \\ \hline y=-1\pm\sqrt{5} \end{array}$$

$$y=-1+\sqrt{5} \text{ or } y=-1-\sqrt{5}$$

Example 12: Solve for x: $2x^2 + 3x + 2 = 0$.

Arrange in the form of $ay^2 + by = -c$:

$$2x^2 + 3x = -2$$

Because $a \neq 1$, multiply the equation by $\frac{1}{2}$.

$$x^2 + \frac{3}{2}x = -1$$

Add $\left[\left(\frac{1}{2}\right)\left(\frac{3}{2}\right)\right]^2$, or $\frac{9}{16}$, to both sides.

$$x^2 + \frac{3}{2}x + \frac{9}{16} = -1 + \frac{9}{16}$$

$$x^2 + \frac{3}{2}x + \frac{9}{16} = -\frac{7}{16}$$

$$\left(x + \frac{3}{4}\right)^2 = -\frac{7}{16}$$

Take the square root of both sides.

$$x + \frac{3}{4} = \pm\sqrt{\frac{-7}{16}}$$

$$x + \frac{3}{4} = \pm\frac{\sqrt{-7}}{\sqrt{16}}$$

$$x + \frac{3}{4} = \pm\frac{\sqrt{-7}}{4}$$

$$x = -\frac{3}{4} \pm \frac{\sqrt{-7}}{4}$$

There is no solution in the real number system. It may interest you to know that the completing the square process for solving quadratic equations was used in the equation $ax^2 + bx + c = 0$ to derive the quadratic formula.

Chapter Check-Out

Questions

1. Solve for x: $x^2 + 3x + 2 = 0$.
2. Which of the following is equivalent to $x^2 - 5x = 6$?
   - **A.** $(x - 2)(x - 3) = 0$
   - **B.** $(x + 2)(x + 3) = 0$
   - **C.** $(x - 6)(x + 1) = 0$
   - **D.** $(x - 6)(x - 1) = 0$

3. Solve for x: $x^2 = 5x - 4$.
4. Solve for x: $x^2 - 49 = 0$.
5. Solve for x: $x^2 - 6x = 0$.
6. Solve for x: $3x^2 + 21x = 2x^2 - 3x + 81$.
7. Solve for x by using the quadratic formula: $x^2 + 5x + 2 = 0$.
8. Solve for x by completing the square: $x^2 + 4x + 2 = 0$.
9. **(a)** Find the discriminant for the quadratic equation $x^2 - 3x + 6 = 0$.
   (b) How many real number solutions does the quadratic equation have? Explain your answer.
10. The roots of a quadratic function are $x = 3$ and $x = -4$. Find the factors of the quadratic equation and write the quadratic equation in the form $x^2 + bx + c = 0$.
11. What value of c will make $x^2 - 3x + c$ a perfect square trinomial?

Question 12 refers to the following scenario.

An archer shoots an arrow into the air with a velocity of 25 meters per second. The arrow's height, h, in meters above the ground after t seconds is given by the function $h(t) = -5t^2 + 25t + 1.5$.

12. When will the arrow hit the ground? (Round your answer to the nearest tenth of a second.)

Questions 13–15 refer to the following information.

Which of the following methods would you use to solve the quadratic equations in questions 13–15? Explain why you would apply this method.

(a) factoring

(b) completing the square

(c) square roots

13. $x^2 - 9x + 7 = 0$
14. $x^2 - 16 = 0$
15. $2x^2 + 3x = 2$
16. The length of a rectangle is $2x + 8$ feet and the width is x feet. If the area of the rectangle is 42 square feet, find the dimensions of the rectangle.

Answers

1. $x = -1$ or $x = -2$
2. C
3. $x = 4$ or $x = 1$
4. $x = 7$ or $x = -7$
5. $x = 0$ or $x = 6$
6. $x = 3$ or $x = -27$
7. $x = \dfrac{-5+\sqrt{17}}{2}$ or $x = \dfrac{-5-\sqrt{17}}{2}$
8. $x = -2+\sqrt{2}$ or $x = -2-\sqrt{2}$
9. (a) The discriminant is $\sqrt{b^2 - 4ac} = \sqrt{(-3)^2 - 4(1)(6)} = \sqrt{-15}$.
   (b) Since the discriminant is negative, there are no real number solutions.
10. The factors of the quadratic equation are $(x - 3)$ and $(x + 4)$, and the standard form of the quadratic equation is $x^2 + x - 12 = 0$.
11. $c = \dfrac{9}{4}$
12. When the arrow hits the ground, the height will be equal to 0 meters. Substitute $h(t) = 0$ into the quadratic function and solve for t, the time in seconds. Using the quadratic formula,

$$0 = -5t^2 + 25t + 1.5$$

$$t = \frac{-25 \pm \sqrt{25^2 - 4(-5)(1.5)}}{2(-5)}$$

$$t \approx -0.059 \text{ or } t \approx 5.059$$

    Since t represents time, we can eliminate $t \approx -0.059$. The solution is 5.1 seconds (rounded to the nearest tenth of a second), which means the arrow hits the ground 5.1 seconds after it is shot up in the air.

13. Method (b), completing the square, since the quadratic equation is not easily factorable and the coefficient of x^2 is 1
14. Method (c), square roots, since the quadratic equation is in the form $x^2 + c = 0$

15. Method (a), factoring, since the quadratic equation $2x^2 + 3x - 2 = 0$ is factorable

16. Since the area of a rectangle is $A = lw$, $x(2x + 8) = 42$. Writing the quadratic equation in standard form and solving for x using factoring by grouping gives us

$$\begin{aligned} 2x^2+8x-42&=0 \\ 2x^2+14x-6x-42&=0 \\ 2x(x+7)-6(x+7)&=0 \\ (2x-6)(x+7)&=0 \\ x=3 \text{ or } \cancel{x=-7} \end{aligned}$$

Since length cannot be negative, $x = 3$, and the dimensions of the rectangle are 3 feet by 14 feet.

Chapter 12
APPLICATION PROBLEMS

Chapter Check-In

- ❑ Problem-solving techniques
- ❑ Key words and phrases
- ❑ Simple and compound interest problems
- ❑ Ratio, proportion, and percent problems
- ❑ Number problems
- ❑ Absolute value and inequality problems
- ❑ Age, motion, and coin problems
- ❑ Mixture and work problems

Common Core Mathematics application problems are often the nemesis of even the best math student. For many, the difficulty is not in the computation. The issues stem from reading the information that is given in the English words, translating what is being asked, and applying the best mathematical operations to solve the problems. Becoming proficient in problem-solving application techniques will help you organize and make sense of algebraic problems that arise in everyday life.

Common Core Standard: Seeing Structure in Expressions

Interpret parts of an expression, such as terms, factors, and coefficients (A.SSE.1.a). Interpret complicated expressions by viewing one or more of their parts as a single entity (A.SSE.1.b).

Common Core Standard: Creating Equations

Create equations and inequalities in one variable and use them to solve problems. Include equations arising from linear and quadratic functions, and simple rational and exponential functions (A.CED.1). Create equations in two or more variables to represent relationships between quantities; graph equations on coordinate axes with labels and scales (A.CED.2). Represent constraints by equations or inequalities and by systems of equations and/or inequalities, and interpret solutions as viable or non-viable options in a modeling context (A.CED.3). Rearrange formulas to highlight a quantity of interest using the same reasoning as in solving equations (A.CED.4).

Common Core Standard: Reasoning with Equations and Inequalities

Explain each step in solving a simple equation as following from the equality of numbers asserted at the previous step, starting from the assumption that the original equation has a solution. Construct a viable argument to justify a solution method (A.REI.1). Solve linear equations and inequalities in one variable, including equations with coefficients represented by letters (A.REI.3).

Common Core Standard: Quantities

Choose a level of accuracy appropriate to limitations on measurement when reporting quantities (N.Q.3).

Common Core Standard: Reasoning with Equations and Inequalities

Solve quadratic equations in one variable (A.REI.4). Use the method of completing the square to transform any quadratic equation in x into an equation of the form $(x - p)^2 = q$ that has the same solutions. Derive the quadratic formula from this form (A.REI.4.a). Solve quadratic equations by inspection, taking square roots, completing the square, using the quadratic formula, and factoring as appropriate to the initial form of the equation. Recognize when the quadratic formula gives complex solutions and write them as $a \pm bi$ for real numbers a and b (A.REI.4.b). Solve systems of linear equations exactly and approximately, focusing on pairs of linear equations in two variables (A.REI.6).

Problem-Solving Techniques

One of the main goals of Common Core Mathematics is to reason abstractly as you apply algebraic operations to real-life events. Real-life application problems involve arithmetic, algebra, geometry, statistics, and their combinations. Application problems may initially challenge your critical thinking processes, but once you learn key problem-solving techniques you should be proficient at presenting reasonable proof and connections between the algebraic expressions and narrative scenarios.

Here are five systematic techniques for solving application problems.

1. **Identify what is being asked.** What are you ultimately trying to find? How far a car has traveled? How fast a plane flies? How many items can be purchased?
2. **Underline or write down information you are given in the problem.** Whatever is being asked, find it and then *write it down* or *circle it.* This helps ensure that you are solving for what is being asked and helps point you to a relationship or equation. Note any key words that signal an operation (see the following section, "Key Words and Phrases").
3. **Use visual organizers.** Draw a picture if you can. It is often helpful to draw a diagram, chart, or symbol to help point you in the right direction. Visual pictures organize your thoughts and provide a clear picture about the data compiled so that you can draw conclusions about the problem.
4. **If you can, define a variable and set up an equation or some straightforward system with the given information.** If you're uncertain about how to begin, look at the words around any unfamiliar term and decide the *context* of the problem. Be sensitive to what each problem is asking and look for verbal expressions that signal a math operation. For example, What time? How many? How much? How far? How old? What length?

   Remember to always follow the rules of the order of operations when setting up the equation (see "Order of operations" in Chapter 1, "Foundations for Understanding Algebra"). Follow the rules of the order of operations if two or more of the following are contained in one problem: multiplication, division, exponents, addition, subtraction, or parenthetical (grouping symbols).

5. **Carefully solve the equation or work the necessary computation.** Be sure you are working in the same units (for example, you may have to change feet into inches or pounds into ounces in order to keep everything consistent).

   When solving the equation, ask yourself, "Is all the given information necessary to solve the problem?" Occasionally, you may be given more than enough information to solve a problem. Choose what you need and don't spend needless energy on irrelevant information.

   And finally, double-check to make sure your answer is reasonable. Check to make sure (a) that you don't have an error in computation, (b) that you did not make a mistake in setting up your equation, and (c) that you have answered "all parts" of the problem. One of the most common errors in answering application word problems is the failure to answer what was actually being asked!

Key Words and Phrases

Common Core Mathematics emphasizes your ability to use logic and reasoning as you translate English words that describe a quantitative situation or scenario. In working with application problems, there are some words or phrases that give clues as to how the problem should be solved. These words or phrases allow you to create mathematical expressions that involve symbols and variables.

Common Math Operational Words (Phrases)

Math Operation	*Word (Phrase)*	*Verbal Expression Example*
Add	Sum	The sum of 2, 3, and 6 . . .
	Total	The total of the first six payments . . .
	Addition	A recipe calls for the addition of five pints . . .
	Plus	Three liters plus two liters . . .
	Increase	Her hourly pay was increased by $15.
	More than	This week the enrollment was eight more than last week.
	Added to	If you added $3 to the cost . . .

Math Operation	Word (Phrase)	Verbal Expression Example
Subtract	Difference	What is the difference between . . .
	Fewer	There were 15 fewer men than women . . .
	Remainder	How many are left or what quantity remains . . .
	Less than	A number is five less than another number . . .
	Reduced	If the budget was reduced by \$5,000 . . .
	Decreased	If he decreased the speed of his car by 10 miles per hour . . .
	Minus	Some number minus 9 is . . .
	Subtracted from	One number is subtracted from another number . . .
Multiply	Product	The product of 8 and 5 is . . .
	Of	One-half of the group . . .
	Times	Five times as many girls as boys . . .
	At	The cost of 10 yards of material at \$2.70 a yard is . . .
	Twice	Twice the value of some number . . .
Divide	Quotient	The final quotient is . . .
	Divided by	Some number divided by 12 is . . .
	Divided into	The group was divided into . . .
	Ratio	What is the ratio of . . .
	Half	Half the profits are . . . (dividing by 2)

As you work a variety of application problems, you will discover more key words and phrases to help you deconstruct problems.

Simple Interest

Simple interest is a common application of percents that models linear growth. Note that simple interest is interest earned only on the principal amount, and both rate and time are calculated based on a yearly or annual rate.

Example 1: If you deposit \$500 into a savings account that earns 8% simple interest per year, how much interest will you earn in 5 years?

First, circle what you must find—*interest.* Now use the simple interest formula.

$$\text{Interest} = \text{principal} \times \text{rate} \times \text{time}$$

$$I = prt$$

Simply substitute into the formula.

$$I = \$500(.08)5$$
$$I = \$200$$

Example 2: Using the information from Example 1, how long, in years, would it take for your money to double?

If you would like your money to double, set $I =$ \$500, since \$500 + \$500 = \$1,000. Then, solve the equation for t. Substitute into the formula.

$$500 = 500(.08)t$$
$$500 = 40t$$
$$t = 12.5 \text{ years}$$

After 12.5 years, the principal amount that was deposited, \$500, will be doubled, or \$1,000.

Compound Interest

Compound interest is a common application of exponents that model exponential growth. Example 3 takes you through the process of finding interest each year using the simple interest formula, but starting with the balance from the previous year, which demonstrates that compound interest is interest earned on both the principal amount and the interest already accrued in the account.

You can also use the following formula to find the balance of an account that earns compound interest,

$$A = P\left(1+\frac{r}{n}\right)^{nt}$$

where A is the ending amount, P is the principal or starting amount, r is the annual interest rate expressed as a decimal, n is the number of times the interest is compounded each year, and t is the time in years. This formula is used in Example 4.

Example 3: What will be the final total amount of money after 3 years on an original investment of $1,000 if a 12% annual interest rate is compounded yearly?

First, circle what you must find—*final total amount of money.* Note also that interest will be *compounded yearly.* Therefore, the solution has three parts, one for each year.

Total for first year: Interest = principal × rate × time

$$I = prt$$
$$I = \$1{,}000(.12)1$$
$$I = \$120$$

Thus, the total after 1 year is $1,000 + $120 = $1,120.

Total for second year:

$$I = prt$$
$$I = \$1{,}120(.12)1$$
$$I = \$134.40$$

Note that the principal at the beginning of the second year was $1,120. Thus, the total after 2 years is $1,120 + $134.40 = $1,254.40.

Total for third year:

$$I = prt$$
$$I = \$1{,}254.40(.12)1$$
$$I = \$150.53$$

Note that the principal at the beginning of the third year was $1,254.40. Thus, the total after 3 years is $1,254.40 + $150.53 = $1,404.93.

Example 4: If you deposit $1,200 into an account that earns 5% interest compounded quarterly for 10 years, how much will you have in your account? How much interest was earned?

First, circle what you must find, the *total amount in your account* after 10 years and the *interest earned.* Since interest is compounded quarterly, $n = 4$. So,

$$A = P\left(1+\frac{r}{n}\right)^{nt}$$

$$A = 1,200\left(1+\frac{0.05}{4}\right)^{10\cdot 4}$$

$$A = \$1,972.34$$

The total amount in your account after 10 years is $1,972.34, and the amount of interest earned is $1,972.34 − $1,200 = $772.34.

Ratio and Proportion

Example 5: If Hunter can type 600 pages of manuscript in 21 days, how many days will it take him to type 230 pages if he works at the same rate?

First, circle what you're asked to find—*how many days.* One simple way to work this problem is to set up a "framework" (proportion) using the categories given in the equation. Here, the categories are *pages* and *days.*

Therefore, a framework may be

$$\frac{\text{pages}}{\text{days}} = \frac{\text{pages}}{\text{days}}$$

Note that you also may have used

$$\frac{\text{days}}{\text{pages}} = \frac{\text{days}}{\text{pages}}$$

The answer will be the same as long as the numerators and denominators have the same units. Now simply substitute into the equation for each instance.

$$\frac{600}{21} = \frac{230}{x}$$

Using the cross-products property,

$$600x = 21(230)$$
$$600x = 4{,}830$$
$$\frac{600x}{600} = \frac{4{,}830}{600}$$
$$x = 8\frac{1}{20} \text{ or } 8.05$$

Therefore, it will take $8\frac{1}{20}$ or 8.05 days to type 230 pages. (You also could have simplified the original proportion before solving.)

Percent

Example 6: Thirty students are awarded doctoral degrees at the graduate school, and this number comprises 40% of the total graduate student body. How many graduate students were enrolled?

First, circle what you must find in the problem—*how many graduate students.* Now, in order to substitute into the percent proportion

$$\frac{\text{is}}{\text{of}} = \frac{\%}{100}$$

try rephrasing the question as a simple sentence. For example, in this case,

30 is 40% of what number?

Notice that the 30 is followed by the word *is*; therefore, 30 is the "is" number. 40 is the percent. Notice that *what number* follows the word *of.* Therefore, substituting into the equation,

$$\frac{\text{is}}{\text{of}} = \frac{\%}{100}$$
$$\frac{30}{x} = \frac{40}{100}$$

Using the cross-products property,

$$40x = 3{,}000$$
$$\frac{40x}{40} = \frac{3{,}000}{40}$$
$$x = 75$$

Therefore, the total graduate enrollment was 75 students.

Example 7: Alex is hiking the Appalachian Trail, which is approximately 2,180 miles long. If he walks an average of 16 miles per day for 30 days, what percent of the trail will he complete? Round your answer to the nearest percent.

First, circle what you must find in the problem, the *percent of the trail* Alex will complete. Since Alex hikes 16 miles per day for 30 days, he will travel 480 miles in 30 days.

Using percent proportion and the cross-products property,

$$\frac{x}{100} = \frac{480}{2{,}180}$$
$$2{,}180x = 480(100)$$
$$\frac{\cancel{2{,}180}x}{\cancel{2{,}180}} = \frac{48{,}000}{2{,}180}$$
$$x = 22\frac{2}{109} \approx 22\%$$

Alex will complete 22% of the Appalachian Trail in 30 days.

Percent Change

Percent change is the percent an amount increases or decreases. For example, if you received a salary raise at your job, you can use the percent change formula to calculate the percent increase in your new salary. Percent increase or percent decrease is found by using the ratio of the amount of change to the original amount. To find the percent change (increase or decrease), use the following formula:

$$\text{percent change} = \frac{\text{change in amounts}}{\text{original amount}}$$

Example 8: Last year, Harold earned \$250 a month at his after-school job. This year, his after-school earnings have increased to \$300 per month. What is the percent increase in his monthly after-school earnings?

First, circle what you're looking for—*percent increase.*

Using the formula,

$$\text{percent change} = \frac{\$300 - \$250}{\$250}$$

$$= \frac{\$50}{\$250}$$

$$= \frac{1}{5} = 0.20 = 20\%$$

The percent increase in Harold's after-school salary is 20%.

Number Problems

Example 9: When 6 times a number is increased by 4, the result is 40. Find the number.

First, circle what you must find—*the number.* Letting x stand for the number gives the equation

$$6x + 4 = 40$$

Subtracting 4 from each side gives

$$6x = 36$$

Dividing by 6 gives

$$x = 6$$

So, the number is 6.

Example 10: One number exceeds another number by 5. If the sum of the two numbers is 39, find the smaller number.

First, circle what you are looking for—*the smaller number.* Now, let the smaller number equal x. Therefore, the larger number equals $x + 5$. Now, use the problem to set up an equation.

$$\underbrace{\text{If the sum of the two numbers}}_{x+(x+5)} \quad \underbrace{\text{is}}_{=} \quad \underbrace{39}_{39}\ldots$$

$$2x+5=39$$

$$2x+5-5=39-5$$

$$2x=34$$

$$\frac{2x}{2}=\frac{34}{2}$$

$$x=17$$

Therefore, the smaller number is 17.

Example 11: If one number is three times as large as another number and the smaller number is increased by 19, the result is 6 less than twice the larger number. What is the larger number?

First, circle what you must find—*the larger number.* Let the smaller number equal x. Therefore, the larger number will be $3x$. Now, using the problem, set up an equation.

$$\underbrace{\text{The smaller number}}_{x} \ \underbrace{\text{increased by}}_{+} \ \underbrace{19}_{19} \ \underbrace{\text{is}}_{=} \ \underbrace{\text{6 less than twice the larger number.}}_{2(3x)-6}$$

$$x+19=6x-6$$

$$-x+x+19=-x+6x-6$$

$$19=5x-6$$

$$19+6=5x-6+6$$

$$25=5x$$

$$5=x$$

Therefore, the larger number, $3x$, is $3(5)$, or 15.

Example 12: The sum of three consecutive integers is 306. What is the largest integer?

First, circle what you must find—*the largest integer*. Let the smallest integer equal x, let the next integer equal $x + 1$, and let the largest integer equal $x + 2$. Now, use the problem to set up an equation.

$$\underbrace{\text{The sum of the three consecutive integers}}_{x+(x+1)+(x+2)} \quad \underbrace{\text{is}}_{=} \quad \underbrace{306.}_{306}$$

$$3x+3=306$$

$$3x+3-3=306-3$$

$$3x=303$$

$$\frac{3x}{3}=\frac{303}{3}$$

$$x=101$$

Therefore, the largest integer, $x + 2$, $= 101 + 2 = 103$.

Absolute Value and Inequality Problems

Example 13: John is a 1,600-meter track runner. He has already competed in three of four races with finishing times of 5 minutes 6 seconds, 5 minutes 15 seconds, and 4 minutes 48 seconds. If John expects to have an average finishing time of at most 5 minutes for the four races, what is the finishing time that John will need for the fourth race to achieve his goal?

First, circle what you are looking for—*the finishing time for the fourth race*. Let the finishing time for the fourth race equal x. Converting seconds to minutes $\left(\frac{\text{seconds}}{60}\right)$ and using the mean formula, set up the inequality.

$$\frac{5.1+5.25+4.8+x}{4}\leq 5$$

$$\frac{15.15+x}{4}\leq 5$$

$$15.15+x\leq 20$$

$$x\leq 4.85$$

Note that .85 minutes must be converted back to seconds (0.85 minutes $\times$ 60 = 51 seconds). John needs to finish the fourth race in 4 minutes and 51 seconds to reach his goal of having an average finishing time of 5 minutes.

Example 14: Ronnie works at a furniture store and earns a base salary of \$325 each week. He makes 6% commission on everything he sells. If Ronnie wants to make at least \$550 by the end of the week, what dollar amount of furniture must he sell?

First, circle what you must find—*what dollar amount of furniture must he sell.* Let the dollar amount of furniture sold equal x. Then set up the inequality and solve.

$$325 + 0.06x \geq 550$$
$$0.06x \geq 225$$
$$x \geq 3{,}750$$

Ronnie must sell at least \$3,750 worth of furniture to reach his goal.

Example 15: In Major League Baseball, a baseball needs to be at a minimum circumference of 9 inches and a maximum circumference of 9.25 inches. Write an absolute value equation that represents the minimum and maximum circumference lengths, where x is equal to the circumference length in inches.

The minimum and maximum circumference lengths are solutions to an absolute value equation. Using the number line, find the halfway point between the two solutions.

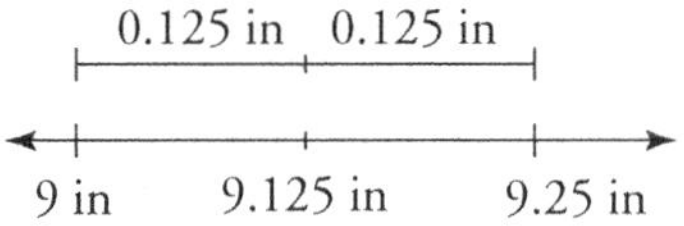

Using the halfway point and the distance from the halfway point to each solution, the absolute value equation is $|x - 9.125| = 0.125$.

Example 16: Organic Granola Cereal Company makes boxes of cereal that should have an ideal weight of 8.9 ounces. A quality control inspector examines and weighs random boxes of cereal taken off the assembly line to check deviations in weight. Any cereal box that varies from the specified weight by more than 2 ounces is taken off the assembly line. What is the allowable range of weights for the box of cereal?

First, circle what you are looking for—*allowable range of weights.* Let the weight in ounces equal w. Since the difference between the actual and ideal weights is at most 2 ounces, use the compound inequality $|w - 8.9| \le 2$.

$$\begin{array}{lcl} w - 8.9 \le 2 & \text{and} & w - 8.9 \ge -2 \\ \underline{\quad +8.9 \quad +8.9} & & \underline{\quad +8.9 \quad +8.9} \\ w \qquad \le 10.9 \text{ oz} & \text{and} & w \qquad \ge 6.9 \text{ oz} \end{array}$$

The weight of a box of cereal must be between 10.9 ounces and 6.9 ounces.

Age Problems

Example 17: Phil is Tom's father. Phil is 35 years old. Three years ago, Phil was four times as old as his son was then. How old is Tom now?

First, circle what you must ultimately find—*how old is Tom now?* Therefore, let t be Tom's age now. Then, 3 years ago, Tom's age would be $t - 3$. Four times Tom's age 3 years ago would be $4(t - 3)$. Phil's age 3 years ago would be $35 - 3 = 32$. A simple chart may also be helpful.

	Now	***3 Years Ago***
Phil	35	32
Tom	t	$t-3$

Now, use the problem to set up an equation.

$$\text{Three years ago } \underbrace{\text{Phil}}_{32} \ \underbrace{\text{was}}_{=} \ \underbrace{\text{four times}}_{4 \text{ times}} \text{ as old as his } \underbrace{\text{son was then.}}_{(t-3)}$$

$$32 = 4(t-3)$$

$$\frac{32}{4} = \frac{4(t-3)}{4}$$

$$8 = t - 3$$

$$8 + 3 = t - 3 + 3$$

$$11 = t$$

Therefore, Tom is now 11 years old.

Example 18: Lisa is 16 years younger than Kathy. If the sum of their ages is 30, how old is Lisa?

First, circle what you must find—*how old is Lisa?* Let Lisa's age equal x. Therefore, Kathy's age is $x + 16$. (Note that since Lisa is 16 years *younger* than Kathy, you must *add* 16 years to Lisa's age to denote Kathy's age.) Now, use the problem to set up an equation.

$$\text{If } \underbrace{\text{the sum of their ages}}_{\text{Lisa's age + Kathy's age}} \ \underbrace{\text{is}}_{=} \ \underbrace{30,}_{30} \text{ how old is Lisa?}$$

$$x+(x+16)=30$$
$$2x+16=30$$
$$2x+16-16=30-16$$
$$2x=14$$
$$\frac{2x}{2}=\frac{14}{2}$$
$$x=7$$

Therefore, Lisa is 7 years old.

Motion Problems

Motion problems are also known as *distance* problems. These types of problems calculate the distance, rate of speed, and time that it takes to travel from one point to another.

Example 19: How long will it take a bus traveling 72 km/hr to go 36 km?

First, circle what you're trying to find—*how long will it take* (time). Motion problems are solved by using the following formula:

$$\text{distance} = \text{rate} \times \text{time}$$
$$d = rt$$

Therefore, simply substitute 72 km/hr for the rate (or speed) of the bus and 36 km for the distance.

$$d = rt$$
$$36 \text{ km} = (72 \text{ km/hr})(t)$$
$$\frac{36}{72} = \frac{72t}{72}$$
$$\frac{1}{2} = t$$

Therefore, it will take one half-hour for the bus to travel 36 km at 72 km/hr.

Example 20: How fast, in miles per hour, must a car travel to go 600 miles in 15 hours?

First, circle what you must find—*how fast* (rate). Now, using the equation $d = rt$, simply substitute 600 for the distance and 15 for the time.

$$d = rt$$
$$600 = r(15)$$
$$\frac{600}{15} = \frac{r(15)}{15}$$
$$40 = r$$

So, the rate is 40 miles per hour.

Example 21: Mrs. Benevides leaves Burbank at 9 a.m. and drives west on the Golden State Freeway at an average speed of 50 miles per hour. Ms. Twill leaves Burbank at 9:30 a.m. and drives west on the Golden State Freeway at an average speed of 60 miles per hour. At what time will Ms. Twill overtake Mrs. Benevides, and how many miles will they each have gone?

First, circle what you are trying to find—*at what time* and *how many miles.* Now, let t stand for the time Ms. Twill drives before overtaking Mrs. Benevides. Then Mrs. Benevides drives for $t + \frac{1}{2}$ hours before being overtaken. Set up the following chart.

	Rate (r)	×	Time (t)	=	Distance (d)
Ms. Twill	60 mph	×	t	=	$60t$
Mrs. Benevides	50 mph	×	$t+\frac{1}{2}$	=	$50\left(t+\frac{1}{2}\right)$

Because each travels the same distance,

$$60t = 50\left(t+\frac{1}{2}\right)$$
$$60t = 50t + 25$$
$$10t = 25$$
$$t = 2.5$$

Ms. Twill will overtake Mrs. Benevides after 2.5 hours of driving. The exact time can be figured out by using Ms. Twill's starting time of 9:30: 9:30 + 2.5 hours = 12 noon. Since Ms. Twill has traveled for 2.5 hours at 60 mph, she has traveled 2.5 × 60, which is 150 miles. So, Mrs. Benevides is overtaken at 12 noon, and each has traveled 150 miles.

Coin Problems

Example 22: Tamar has four more quarters than dimes. If he has a total of $1.70, how many quarters and dimes does he have?

First, circle what you must find—*how many quarters and dimes.* Let x stand for the number of dimes; then $x + 4$ is the number of quarters. Therefore, $0.10x$ is the total value of the dimes and $0.25(x + 4)$ is the total value of the quarters. Setting up the following chart can be helpful.

	Number	Value	Amount of Money
Dimes	x	0.10	$0.10x$
Quarters	$x + 4$	0.25	$0.25(x + 4)$

Now, use the table and problem to set up an equation.

$$0.10x+0.25(x+4)=1.70$$
$$10x+25(x+4)=170$$
$$10x+25x+100=170$$
$$35x+100=170$$
$$35x=70$$
$$x=2$$

So, there are two dimes. Since there are four more quarters, there must be six quarters.

Example 23: Sid has $4.85 in coins. If he has six more nickels than dimes and twice as many quarters as dimes, how many coins of each type does he have?

First, circle what you must find—*how many coins of each type.* Let x stand for the number of dimes. Then $x + 6$ is the number of nickels and $2x$ is the number of quarters. Setting up the following chart can be helpful.

	Number	***Value***	***Amount of Money***
Dimes	x	0.10	$0.10x$
Nickels	$x + 6$	0.05	$0.05(x + 6)$
Quarters	$2x$	0.25	$0.25(2x)$

Now, use the table and problem to set up an equation.

$$0.10x+0.05(x+6)+0.25(2x)=4.85$$
$$10x+5(x+6)+25(2x)=485$$
$$10x+5x+30+50x=485$$
$$65x+30=485$$
$$65x=455$$
$$x=7$$

So, there are seven dimes. Therefore, there are thirteen nickels and fourteen quarters.

Mixture Problems

Example 24: Coffee worth \$1.05 per pound is mixed with coffee worth 85¢ per pound to obtain 20 pounds of a mixture worth 90¢ per pound. How many pounds of each type are used?

First, circle what you are trying to find—*how many pounds of each type.* Let the number of pounds of \$1.05 coffee be denoted as x. Therefore, the number of pounds of 85¢-per-pound coffee must be the remainder of the 20 pounds, or $20 - x$. Make a chart for the cost of each type and the total cost.

	Cost per lb	×	Amount in lbs	=	Total Cost of Each
\$1.05 coffee	\$1.05	×	x	=	\$1.05x
\$0.85 coffee	\$0.85	×	$20 - x$	=	\$0.85 $(20 - x)$
Mixture	\$0.90	×	20	=	\$0.90(20)

Now, set up the equation.

$$\underbrace{\text{total cost of one type}}_{\$1.05x}\ \underbrace{\text{plus}}_{+}\ \underbrace{\text{total cost of other type}}_{\$0.85(20-x)}\ \underbrace{\text{equals}}_{=}\ \underbrace{\text{total cost of mixture}}_{\$0.90(20)}$$

$$1.05x + 17.00 - 0.85x = 18.00$$

$$17.00 + 0.20x = 18.00$$

$$-17.00 + 17.00 + 0.20x = 18.00 - 17.00$$

$$0.20x = 1.00$$

$$\frac{0.20x}{0.20} = \frac{1.00}{0.20}$$

$$x = 5$$

Therefore, 5 pounds of coffee worth \$1.05 per pound are used. And $20 - x$, or $20 - 5$, or 15 pounds of 85¢-per-pound coffee are used.

Example 25: Solution A is 50% hydrochloric acid, while solution B is 75% hydrochloric acid. How many liters of each solution should be used to make 100 liters of a solution that is 60% hydrochloric acid?

First, circle what you're trying to find—*liters of each solution.* Let x stand for the number of liters of solution A. Therefore, the number of liters of solution B must be the remainder of the 100 liters, or $100 - x$. Next, make the following visual chart.

	% of Acid	Liters	Concentration of Acid
Solution A	50%	x	$0.50x$
Solution B	75%	$100 - x$	$0.75(100 - x)$
New solution	60%	100	$0.60(100)$

Now, set up the equation.

$$\begin{aligned} 0.50x + 0.75(100 - x) &= 0.60(100) \\ 0.50x + 75 - 0.75x &= 60 \\ -75 \qquad\qquad &\quad -75 \\ \hline 0.50x \qquad -0.75x &= -15 \\ -0.25x &= -15 \\ \frac{-0.25x}{-0.25} &= \frac{-15}{-0.25} \\ x &= 60 \end{aligned}$$

Therefore, using the chart, 60 liters of solution A are used and $100 - 60 = 40$ liters of solution B are used.

Work Problems

Example 26: Ernie can plow a field alone in 4 hours. It takes Sid 5 hours to plow the same field alone. If they work together (and each has a plow), how long will it take to plow the field?

First, circle what you must find—*how long . . . together.* Work problems of this nature may be solved by using the following equation.

$$\frac{1}{\text{1st person's time}} + \frac{1}{\text{2nd person's time}} + \frac{1}{\text{3rd person's time}} + \text{etc.} = \frac{1}{\text{time together}}$$

Therefore,

$$\frac{1}{\text{Ernie's time}} + \frac{1}{\text{Sid's time}} = \frac{1}{\text{time together}}$$

$$\frac{1}{4} + \frac{1}{5} = \frac{1}{t}$$

Finding a common denominator,

$$\frac{5}{20}+\frac{4}{20}=\frac{1}{t}$$
$$\frac{9}{20}=\frac{1}{t}$$

Using the cross-products property,

$$9t = 20$$
$$\frac{9t}{9}=\frac{20}{9}=2\frac{2}{9}\text{ hours}$$

Therefore, it will take them $2\frac{2}{9}$ hours working together to plow the field.

Number Problems with Two Variables

Example 27: The sum of two numbers is 15. The difference of the same two numbers is 7. What are the two numbers?

First, circle what you're looking for—*the two numbers.* Let x stand for the larger number and y stand for the smaller number. Now, set up two equations.

The sum of the two numbers is 15.

$$x + y = 15$$

The difference is 7.

$$x - y = 7$$

Now, solve using the elimination method.

$$\begin{aligned} x+y&=15 \\ x-y&=\ \ 7 \\ \hline 2x\qquad&=22 \\ x\qquad&=11 \end{aligned}$$

Substituting $x = 11$ into the first equation gives

$$11 + y = 15$$
$$y = 4$$

So, the numbers are 11 and 4.

Example 28: The sum of twice one number and three times another number is 23 and their product is 20. Find the numbers.

First, circle what you must find—*the numbers.* Let x stand for the number being multiplied by 2 and y stand for the number being multiplied by 3.

Now set up two equations.

The sum of twice a number and three times another number is 23.

$$2x + 3y = 23$$

Their product is 20.

$$x(y) = 20$$

Rearranging the first equation gives

$$3y = 23 - 2x$$

Dividing each side of the equation by 3 gives

$$y = \frac{23}{3} - \frac{2x}{3}$$

Substituting the first equation into the second gives

$$x\left(\frac{23}{3} - \frac{2x}{3}\right) = 20$$
$$\frac{23x}{3} - \frac{2x^2}{3} = 20$$

Multiplying each side of the equation by 3 gives

$$23x - 2x^2 = 60$$

Rewriting this equation in standard quadratic form gives

$$2x^2 - 23x + 60 = 0$$

Solving this quadratic equation using factoring gives

$$(2x - 15)(x - 4) = 0$$

Setting each factor equal to 0 and solving gives

$$\begin{aligned} 2x - 15 = 0 &\text{ or } x - 4 = 0 \\ 2x = 15 &\text{ or } \quad x = 4 \\ x = \frac{15}{2} &\text{ or } \quad x = 4 \end{aligned}$$

With each x value we can find its corresponding y value.

If $x = \frac{15}{2}$, then $y = \frac{23}{3} - \frac{2\left(\frac{15}{2}\right)}{3}$ or $y = \frac{23}{3} - \frac{15}{3} = \frac{8}{3}$.

If $x = 4$, then $y = \frac{23}{3} - \frac{2(4)}{3}$ or $y = \frac{23}{3} - \frac{8}{3} = \frac{15}{3} = 5$.

Therefore, this problem has two sets of solutions. The number being multiplied by 2 is $\frac{15}{2}$ and the number being multiplied by 3 is $\frac{8}{3}$, or the number being multiplied by 2 is 4 and the number being multiplied by 3 is 5.

Example 29: Bamboo plants are the fastest growing plants in the world and are also the main source of food for giant pandas. Many species of bamboo are grown at the San Diego Zoo to feed the giant pandas. Suppose the zoo staff plants a certain species of bamboo already 20 inches tall that grows at an average rate of 10 inches per day. The staff then plants another species of bamboo already 5 inches tall that grows at an average rate of 15 inches per day. How many days will it take the second species of bamboo to catch up to the first species? How tall will each species of bamboo be?

There are two things you must find—*the number of days it will take the second species to catch up to the first species* and *the height of each plant at that point.* Let t be the number of days and $h(t)$ be the height of the bamboo plant in inches. Set up a system of equations and solve for t and $h(t)$.

$$h(t) = 20 + 10t$$
$$h(t) = 5 + 15t$$

Using substitution,

$$5 + 15t = 20 + 10t$$
$$5 + 5t = 20$$
$$5t = 15$$
$$t = 3$$

Using the first equation, substitute $t = 3$ into the equation to find the height in inches.

$$h(t) = 20 + 10t$$
$$h(3) = 20 + 10(3)$$
$$h(3) = 50$$

In 3 days, the height of the second species will catch up to the height of the first species at 50 inches.

Chapter Check-Out

Questions

1. Find the total interest on $140 at a 5% annual rate for 2 years if the interest is compounded annually.
2. Your friend borrows $1,500 from you and promises to give the money back in 2 years with 4% interest. How much money will you receive at the end of the second year?
3. You deposit $15,000 into a savings account that earns a simple interest rate of 1.25% each year. How long was the money in your account if you earned a total of $937.50 in simple interest?

4. The scale of a map is 1 inch : 100 miles. If the distance from Flagstaff, Arizona, to Phoenix, Arizona, is 2.8 inches on the map, how many miles apart are the two cities?

5. Last week, the price of an avocado at the supermarket was \$0.99. This week, the price is \$1.50 for two avocados. By what percent increase did the price of an avocado go up? Round your answer to the nearest percent.

6. James is 6 years older than Simin. In 2 years the sum of their ages will be 20. How old is Simin now?

7. Jessica is making potato salad for a family reunion. If 4 gallons of potato salad are needed to feed 100 people, find the number of gallons of potato salad needed to feed the 140 family members going the reunion.

8. Mason is selling his house and his Realtor earns a 6% commission based on the sales price of the home. If Mason's Realtor earned \$32,100 in commission, how much did Mason's house sell for?

9. An adult's normal body temperature can be anywhere between 98.1 and 99.1 degrees Fahrenheit. Write an absolute value equation that represents the maximum and minimum body temperatures.

10. Starting from 500 feet away, your friend is kayaking toward you at a constant speed of 3.5 feet per second. The distance d (in feet) your friend is from you after t seconds is given by the absolute value equation $d = |500 - 3.5t|$. At what times (in minutes) will your friend be 50 feet away from you? Round to the nearest tenth of a minute.

11. Ellen has collected nickels and dimes worth a total of \$6.30. If she has collected 70 coins in all and each is worth its face value, how many of each kind does she have?

12. Terrell can put up a wood fence in 5 hours if he works alone. It takes Miri 6 hours to put up the same wood fence if she works alone. If they work together, how long will it take them to put up the same wood fence?

13. The sum of two numbers is 40 and their product is 300. What are the two numbers?

14. A fishing boat travels upstream from Lees Ferry to Glen Canyon Dam for 3 hours. The return trip takes only 2.5 hours because the boat is traveling 1.1 miles per hour faster with the downstream current. How far, in miles, does the boat travel upstream?

15. Katie sold two different types of cookies to raise money for a school fundraiser. The peanut butter cookies sold at $7 per box and the mint cookies sold at $5 per box. Katie made $140 over the weekend, but she did not keep track of how many of each box she sold. If she started with four more boxes of mint than peanut butter cookies initially, how many of each type of cookie did Katie sell?

Answers

1. $14.35
2. $1,620
3. 5 years
4. 280 miles
5. 52% increase
6. 5 years old
7. 5.6 gallons
8. $535,000
9. $|x - 98.6| = 0.5$
10. 2.1 minutes and 2.6 minutes
11. 14 nickels and 56 dimes
12. $2\frac{8}{11}$ hours
13. 10 and 30
14. 16.5 miles
15. 10 boxes of peanut butter cookies and 14 boxes of mint cookies

REVIEW QUESTIONS

Use these review questions to practice what you've learned in this book. After you work through the review questions, you'll be well on your way to understanding the basic concepts of Algebra I.

Questions

Chapter 1

1. Which of the following are rational numbers? $0,\ 2,\ \sqrt{5},\ \frac{1}{2},\ 0.6$

2. Which of the following are prime numbers? 2, 6, 9, 11, 15, 17

3. The multiplicative inverse of $\frac{2}{3}$ is ______.

For each of the following statements in 4–7, write True or False.

4. 1 is a prime number.

5. $\sqrt{5}$ is an irrational number.

6. $\frac{2}{3}$ is a rational number.

7. -1 is a natural number.

8. Which choice best illustrates the associative property of addition?

   A. $(ab) + c = (ba) + c$
   B. $(a + b) + c = a + (b + c)$
   C. $a(b + c) = ab + ac$
   D. $a + b = b + a$

9. Rewrite 9 in exponential notation using 3 as a base.

10. Simplify $\frac{2^5}{2^{-3}}$. Write your answer in exponential notation.

11. Your friend says that $7^2 \times 7^5$ is equivalent to 49^7. Explain and correct the error that was made.

12. Find the value of n that will make the number sentence below a true statement.

$$\frac{8^6}{4^6} = \frac{8^6}{8^n} = 8^2$$

13. Rewrite $(3^4)^5$ as an equivalent expression in exponential notation.

14. Simplify the expression $\left(\frac{2}{3}\right)^2 \cdot \left(\frac{1}{4}\right)^0$.

15. The expression $5^3 \times 2^4$ is equivalent to which of the following numerical expressions?

A. $(5 \times 2)^7$

B. 125×16

C. 10^{12}

D. 15×8

16. Write an inequality that compares the following two roots.

$$-\sqrt{64} \qquad -\sqrt[3]{125}$$

17. Find an approximation of $\sqrt{18}$ to the nearest tenth.

18. Is the sum of 2.5 and $\sqrt{3}$ rational or irrational? Explain.

19. Describe and correct the error in simplifying the radical.

$$\sqrt{40} = \sqrt{4 \cdot 10} = 4\sqrt{10}$$

20. Simplify: $2[(3^2 + 4) + 2(1 + 2)]$.

21. The number 12,120 is divisible by which numbers between 1 and 10?

Question 22 refers to the following statement.

The sum of two numbers is −3.

22. Yes or no: Can both numbers be positive?

If not, explain what you notice about the numbers and write down three different numerical expressions that give a sum of −3.

23. Marine biologists in a submarine descend to the bottom of the ocean at a rate of 5 feet every 4 seconds. If the submarine reaches the bottom of the ocean in 13 minutes, what is the depth of the submarine?

24. The product of an odd number of negative factors is __________.

25. Simplify: $\frac{3}{5}+\frac{6}{7}\cdot\frac{14}{9}$.

26. A recipe calls for $2\frac{1}{4}$ cups of sugar. If you need to increase the recipe $1\frac{1}{2}$ times, how much sugar do you need?

27. A furniture upholsterer needs $3\frac{1}{3}$ yards of fabric to re-upholster a chair. How many chairs can be covered with $20\frac{2}{3}$ yards of fabric?

28. Denver, Colorado, is referred to as the "Mile-High City," with an elevation of 5,280 feet above sea level. New Orleans, Louisiana, has an elevation of approximately 7 feet below sea level. What is the difference of their elevations?

29. Approximately 7% of a person's total body weight is blood. A pint of blood weighs approximately 1 pound. A man who weighs 142 pounds has how many pints of blood?

30. Change $0.\overline{4}$ to a fraction.

31. 20 is what percent of 400?

32. Terry bought a computer for 25% off the regular price of $1,950. How much did the computer cost with 9% sales tax?

33. A car is purchased for x dollars and depreciates by about 13% of the original value each year for the next 3 years. Write an expression to describe the approximate value of the car after 3 years.

34. Explain why 0.423×10^5 is not written in scientific notation.

For questions 35 and 36, express each answer in scientific notation.

35. $(4 \times 10^3)(3 \times 10^2) =$

36. $(8 \times 10^{-2}) \div (2 \times 10^4) =$

Chapter 2

For questions 37–40, use the following information.

- $U = \{p \in \mathbb{Z} | 0 \leq p \leq 10\}$ is the universal set,
- $A = \{r | r \text{ is an even number}, r < 10\}$, and
- $B = \{x | x \text{ is a factor of } 12\}$

37. Write sets A and B in roster form.

38. Find A'.

39. Find $A \cap B$.

40. Find $A \cup B$.

41. True or false: {Tom, Bob, Sam} ~ {1, 2, 3}.

For questions 42–45, write an algebraic expression for each word phrase.

42. Five less than twice a number, n

43. One-half times the sum of a number, n, and 15

44. The quotient of a number, x, and 3, increased by the product of 5 and another number y

45. Explain why sets R and M are disjoint.

$$R = \{\text{triangle, square, rectangle}\}$$

$$M = \{\text{cube, rectangular prism, pyramid}\}$$

46. Evaluate $2x + 4y^2$ if $x = 3$ and $y = -2$.

47. Peter writes an equation $h = 6c + 2$. If the value of c is increased by 5, what will happen to the value of h?

48. Jorge's family takes a trip to an amusement park in Montana. It costs \$10 to park. Admission is \$14 for each adult and \$7 for children under 12 years old. If the total cost is \$80 (including parking), write an algebraic expression that gives the total cost for the trip to the amusement park for 2 adults and x children, where x represents the number of children. How many children under 12 years old went to the amusement park?

49. Leo is painting seven cubes for the school play. Each cube has a side length of 1.5 feet. Use the expression $6s^2$ to find the surface area of a cube, where s is the side length of the cube. If 1 gallon of paint covers 350 ft^2, what fraction of a gallon of paint will Leo need to paint all seven cubes for the school play?

Chapter 3

50. State the properties of equality that are used to solve the equation $\frac{y}{8} - 3 = 9$. Which property of equality is applied first? Which property of equality is applied second?

51. Which of the following equations is not equivalent to $5r + 7 = 3r - 15$?

A. $2r + 32 = 10$

B. $3r + 12 + 2r = -43$

C. $-4(2r + 8) = 56$

D. $\frac{r+9}{10} = -2$

52. Your friend tells you that the equation $-3(2 - x) = 3x + 6$ is an identity. Is your friend correct? Explain your reasoning.

53. Solve for x in terms of a and c: $2x - ax = 3c$.

54. Solve for c: $\frac{a}{c} = \frac{b}{d}$.

55. Solve the proportion for x: $\frac{2}{2x+7} = \frac{5}{4x}$.

56. Describe and correct the errors in solving the following equation:

$$2x+4-3x=5(x+8)$$
$$2x+4-3x=5x+8$$
$$5x+4=5x+8$$
$$4\neq 8$$

The equation has no solution.

Use the following information for questions 57 and 58.

The surface area of a right cone can be calculated using the formula $S.A. = \pi r\ell + \pi r^2$, where r is the radius and ℓ is the slant height of the right cone.

57. Write an expression that can be used to represent the slant height, ℓ.

58. Find the slant height of a cone with a surface area of 8π cm^2 and radius of 2 cm.

Chapter 4

59. Solve for x and y:

$$3x+2y=1$$
$$2x-3y=-8$$

60. Solve for m and n:

$$m=n+3$$
$$m+2n=9$$

61. Can a system of linear equations have more than one solution? Explain your answer.

62. Determine whether $(1, -2)$ is a solution to the following system of equations. Explain your answer.

$$2x+3y=-4$$
$$4x-y=2$$

63. For what value of a will the system of equations have no solution?

$$ax+4y=-4$$
$$2x+8y=-10$$

For questions 64–66, write down whether the system of equations has one solution, no solution, or infinitely many solutions when solving by substitution.

64. $3 = 8$

65. $2x = 4$

66. $5 = 5$

67. What can be deduced about the solution of a system of linear equations that has the same slope but different y-intercepts?

Chapter 5

68. Simplify: $2xy^2 + xy^2 - 6xy^2$. Write the degree of the simplified monomial.

69. Simplify: $(3x^4y)^3(-5x^2y)$.

70. Simplify: $\dfrac{9x^3y^6z^2}{3x^2y^5z}$.

71. Write $x^{-5}yz^{-2}$ in simplest form.

72. What is the result when $(6a^2 - 3b + 8)$ is subtracted from $(8a^2 - 4b - 5)$?

73. Explain why $4x^2y$ and $-3xy^2$ are not like terms.

74. The length of a rectangle is $(5x + 2y)$ inches and the width is $(2x + 3y)$ inches. Write an expression for the perimeter of the rectangle in its simplest form.

75. Write the polynomial that represents the area of the shaded region of the figure below in standard form.

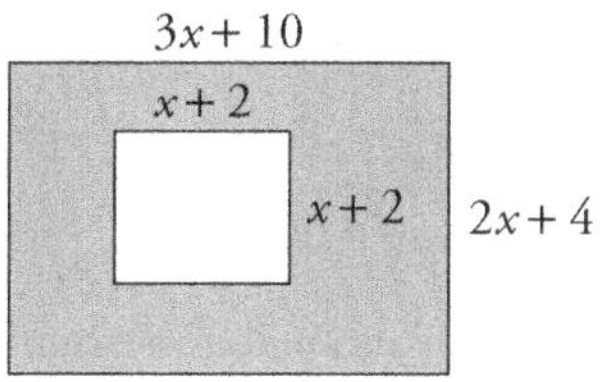

76. Simplify: $\dfrac{12m^2n^2 + 14mn^2}{2mn}$.

77. The area of a rectangular frame is $(x^2 - 2x - 24)$ square inches and the width is $(x + 4)$ inches. Write an expression for the length of the frame.

78. Factor: $2y^2 - 8y$.

79. Factor: $9x^2 - 16$.

80. Find two different coefficients for the x term in the box below so that the trinomial can be factored into a product of two binomials. Then factor each trinomial.

$$x^2 - \square x + 36$$

81. The volume of a rectangular prism is $(4a^3 + 6a^2 + 2a)$ cubic feet. Find the dimensions of the rectangular prism in terms of a.

82. Factor: $m^2 + 5mn + 4n^2$.

83. Factor: $a + 6 + ab + 6b$.

Chapter 6

84. Simplify: $\dfrac{5x^6}{15x^4}$.

85. Simplify: $\dfrac{x^2-3x+2}{3x-6}$.

86. Find the product of $\dfrac{x^3}{2y} \cdot \dfrac{3y^2}{4x}$.

87. Find the product of $\dfrac{x^2+2x+1}{x+2} \cdot \dfrac{2x+4}{x+1}$. State any excluded values.

88. Find the quotient of $\dfrac{6x^3}{7} \div \dfrac{2x^2}{y}$.

89. Find the quotient of $\dfrac{6y+12}{8} \div \dfrac{y+2}{4}$.

90. What error was made in the rational expression below?

$$\frac{3x-2}{x+1} - \frac{2x-1}{x+1} = \frac{\cancel{x}-1}{\cancel{x}+1} = -1$$

91. The expression $\frac{x-2}{2-x}$ is simplified because the numerator and denominator have no common factors other than 1. Is this statement correct? Explain your answer.

92. Explain why $\frac{2\sqrt{x}-5}{x^2+2x-1}$ is not a rational expression.

93. Find the sum of $\frac{5}{x}+\frac{7}{y}$.

94. Find the sum of $\frac{3}{a^3b^3}+\frac{2}{a^4b^2}$.

95. Find the difference of $\frac{3x}{x-3}-\frac{2x}{x+1}$.

96. Find the sum of $\frac{x}{x^2-16}+\frac{4x}{x^2+5x+4}$.

97. A tour boat takes tourists round-trip from New Orleans, Louisiana, to Baton Rouge, Louisiana, on the Mississippi River. The distance between the cities is 81 miles. The boat travels 25% faster downstream to New Orleans due to the rapid currents. Let r be the speed in miles per hour from New Orleans to Baton Rouge. Using the formula $d = rt$, write a simplified expression for the round-trip travel time on the boat.

Use the following diagram to answer questions 98 and 99.

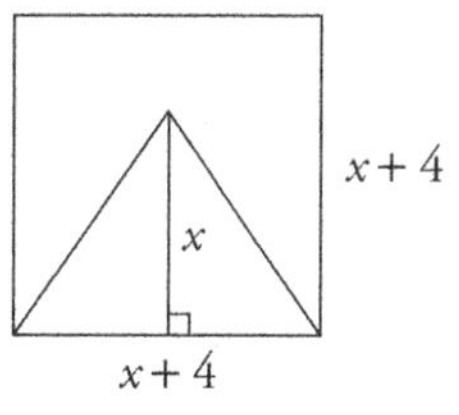

98. Write a simplified expression for the ratio of the area of the triangle to the area of the square.

99. If $x = 2$, what fraction of the area of the square does the triangle represent?

Chapter 7

100. For what value of a will the solution of the following inequality be $x > -2$?

$$ax + 5 < 15$$

101. Solve the inequality $3x + 4 \geq 5x - 8$. Write the solution in interval notation and graph the solution on a number line.

102. Using the Fahrenheit scale, the freezing point of water is 32°F and the boiling point is 212°F. Write this range as a compound inequality.

103. Which set or inequality does not match the graph below?

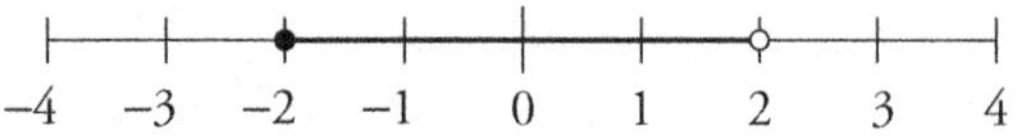

A. $-2 \leq x < 2$

B. $x \geq -2$ and $x < 2$

C. $[-2, 2)$

D. $(-2, 2]$

104. Graph: $\{x \mid x < 6\}$ on a number line.

105. Simplify: $|-6 - 3|$.

106. Write an absolute value equation that has 5 and 8 as solutions.

107. Solve for x: $2|x - 2| + 5 = 13$.

108. Solve for x and graph on a number line: $3|x + 1| + 2 > 8$.

Chapter 8

109. Is $x + \frac{3}{y} = 9$ linear or non-linear? Explain your answer.

110. Graph the equation of a line that passes through the point $(-2, 3)$ and has a slope of 2. Then write an equation of the line in slope-intercept form.

111. Graph: $2y + 4 = 3x$.

112. Find the slope of $2x + y = 6$.

113. Write an equation of a line whose slope is undefined and passes through the point $(-4, 3)$.

114. Is the point $(5, -3)$ a solution to the line $y - 5 = -1(x + 3)$?

115. Which of the following statements is false about the graph of the equation $y + 7 = -3(x - 1)$?

A. The line has a slope of -3.

B. The line passes through the point $(1, -7)$.

C. The line is perpendicular to the graph of the equation $y = \frac{1}{3}x + 10$.

D. The equation is written in slope-intercept form.

116. Find the equation of the line, in slope-intercept form, passing through the point $(5, 3)$ with a slope of 2.

117. Find the equation of the line, in point-slope form, passing through the point $(6, -3)$ with a slope of $\frac{1}{3}$. Then write it in standard form.

118. A linear relationship exists between the Celsius and Fahrenheit temperature scale. Since you know that 0°C or 32°F is the freezing point of water and 100°C or 212°F is the boiling point of water, write an equation that describes the relationship between the two scales, where the Fahrenheit temperature, F, is written as a function of Celsius temperature, C.

119. Find the equation of the line, in standard form, passing through the points $(2, 3)$ and $(-1, -3)$. Then find the x-intercept of the line.

120. An aquarium tank that holds 20 gallons of water is leaking at a rate of 1.5 gallons every 20 minutes. Write an equation that represents the number of gallons, $G(m)$, in terms of the number of minutes, m. If the tank starts to leak at 8 a.m., what time will the tank be empty?

121. The linear function $N(b) = 400{,}000 - 5{,}000b$ represents the number of barrels remaining from an oil field each year, where b is the number of barrels of oil. Interpret the slope and the y-intercept in the context of the situation.

122. The relationship between the side length of a square and the area of the square is displayed in the table below. Is this relationship linear? Explain your answer.

Side Length	***Area***
2	4 ft^2
4	16 ft^2
6	36 ft^2
8	64 ft^2

Question 123 refers to the following information.

The relationship between the length of a rectangle and the perimeter of the rectangle is displayed in the table below.

Length	***Perimeter***
2	20 ft
4	24 ft
5	26 ft
9	34 ft

123. Using the table above, write a linear equation that models the perimeter, $P(x)$, of the rectangle in terms of the length, x.

124. Graph $y \geq x + 1$.

125. Graph $3x - 2y < 9$ using intercepts. Is $(2, -1)$ a solution to the linear inequality? Explain your answer.

Chapter 9

126. Which of the following are graphs of functions?

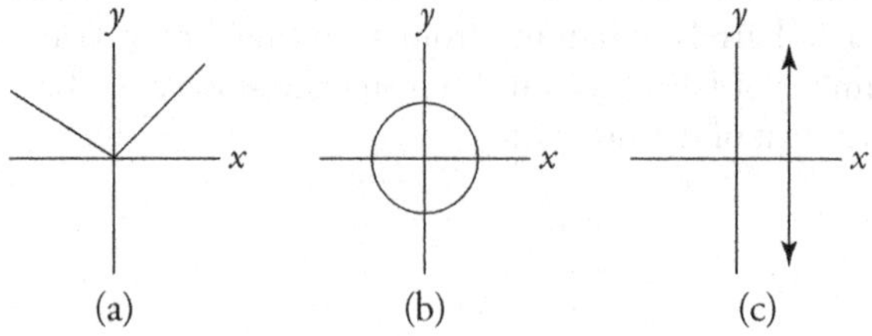

127. Explain why the table below does not represent a function.

x	**y**
−2	4
−1	0
3	−4
−1	5

128. If $f(x) = 3x^2 - 2x - 1$, what is $f(3)$?

129. The function $y = -4x + 8$ represents the weight y (in ounces) of a candy bar remaining after you take x bites. The domain of the function is 0, 1, and 2. What is the range of the function?

130. Which of the following statements is NOT true about the function in the graph below?

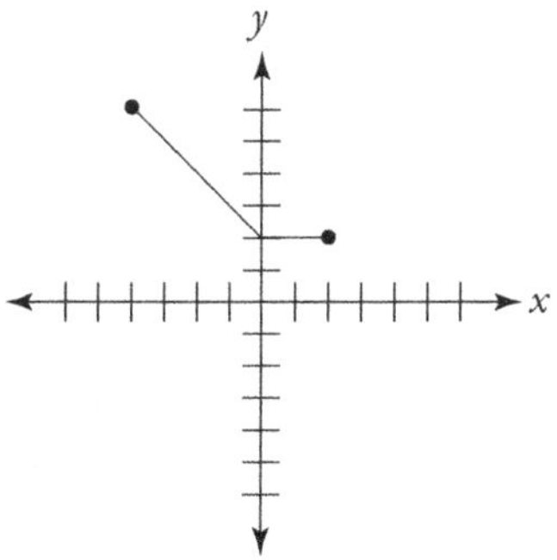

A. The domain is $-4 \leq x \leq 2$.

B. The range is $0 \leq y \leq 6$.

C. The function passes the vertical line test.

131. The temperature of a turkey cooking in the oven can be represented by a function T(m), where the input, m, represents the number of minutes the turkey has been cooking.

Which choice below does T(120) represent?

A. the temperature of the turkey when it was initially placed in the oven

B. the temperature of the turkey after 120 minutes of baking

C. the number of minutes it takes for the turkey to reach 120 degrees

D. the number of minutes it takes to completely cook the turkey

132. Using the graph below, find the value of x if $f(x) = -3$.

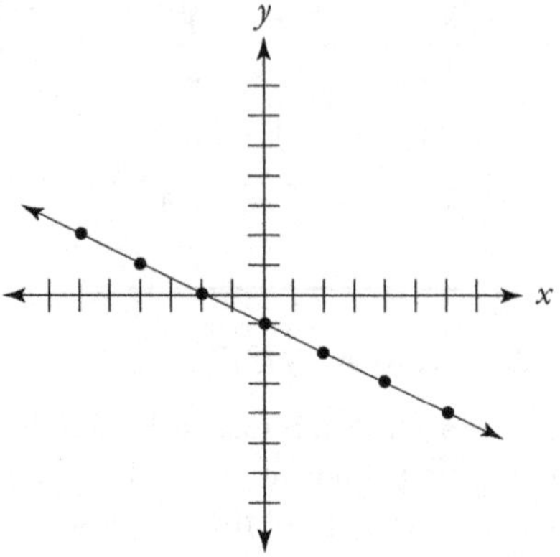

133. If y varies directly with x, find the constant of variation when y is 3 and x is 9. Then write the equation for the direct variation.

134. If y varies inversely with x and the constant of variation is 2, find y when x is 8.

135. If y varies inversely with x and $y = 12$ when $x = 3$, write an equation for the inverse variation.

Questions 136 and 137 refer to the following information.

Determine whether the data in the tables below represent a direct variation or an inverse variation. Find the constant of variation and write an equation for the direct or inverse variation.

136.

x	y
2	14
4	7
8	3.5

137.

x	y
2	−8
3	−12
4	−16

For questions 138 and 139, determine whether the scenario is an example of a direct variation or an indirect variation. Explain your answer.

138. The price of five pizzas at \$15.99 each is split equally among a group of people.

139. The price of oranges at the store is \$0.99 per pound.

140. Johnny and Brian are on a seesaw at the playground and want to keep the seesaw balanced. The weight (in pounds) needed to balance the seesaw varies indirectly with the distance (in feet) each person is from the fulcrum. If Johnny weighs 110 pounds and is 5 feet from the fulcrum, and Brian weighs 90 pounds, approximately how far should Brian be from the fulcrum to balance the seesaw? Round your answer to the nearest tenth of a foot.

Chapter 10

141. Simplify: $\sqrt{75}$.

142. Which of the following is false?

A. $\frac{\sqrt{6}}{\sqrt{2}} = \sqrt{3}$

B. $\sqrt{6} \cdot \sqrt{2} = \sqrt{12}$

C. $\sqrt{6} + \sqrt{2} = \sqrt{8}$

D. $\sqrt{2} + \sqrt{2} = 2\sqrt{2}$

143. Explain why $\sqrt{5}$ and $\sqrt{75}$ are not like radicals.

144. Explain why $\frac{4}{2-\sqrt{3}}$ is not in simplest form. Then write the simplified form of the expression.

145. If each variable is nonnegative, then $\sqrt{25x^5y^8} =$

For questions 146–148, write each expression in simplest form.

146. $\sqrt{50} + 3\sqrt{2} =$

147. $6\sqrt{3t^2} \cdot 2\sqrt{12t^3}$

148. $\left(1-\sqrt{12}\right)^2$

For questions 149 and 150, express answers in simplified form with rationalized denominators.

149. $\frac{\sqrt{7}}{\sqrt{2}} =$

150. $\frac{3+\sqrt{2}}{2-\sqrt{2}} =$

151. A carpentry class is constructing a door in which the length is 4 times the width, w. What is the simplified expression for the diagonal length of the door in terms of w?

Chapter 11

152. Solve for x: $x^2 - 2x = 63$.

153. Solve for x: $x^2 - 81 = 0$.

154. Solve for x: $x^2 + 8x = 0$.

155. Solve for x: $3x^2 + 3x + 2 = 2x^2 + x + 1$.

156. Solve for x using the quadratic formula: $x^2 + 3x + 1 = 0$.

157. Solve for x by completing the square: $x^2 + 7x + 4 = 0$.

158. What value of b will make $x^2 + bx + 9$ a perfect square trinomial?

159. The roots of a quadratic function are $x = \frac{5}{2}$ and $x = 2$. Find the factors of the quadratic equation and write the quadratic equation in standard form.

160. Find the discriminant for the quadratic equation $x^2 - 10x + 25 = 0$. How many real number solutions does the quadratic equation have?

161. A rock is dropped from a bridge that is 150 feet above the ground. How high the rock is above the ground, h, in feet, after t seconds is given by the function $h(t) = -16t^2 + 150$. When will the rock be positioned 10 feet from the ground? Round your answer to the nearest hundredth of a second.

162. A community garden has a rectangular shape. Its area, A, is 100 square feet; its width, w, is $(x + 2)$ feet; and its length, l, is $(2x + 4)$ feet. The city has approved a larger, similar community garden, which will increase the garden area by 30%. Write an equation that could be used to determine the length and width of the larger garden. Find the new dimensions of the rectangular garden. Round your answer to the nearest tenth.

163. Which method would you use to solve the quadratic equation $x^2 + 5x - 2 = 0$? Explain your answer.

Chapter 12

164. If Tim invests $200 at a 10% annual rate for 3 years compounded quarterly, how much money will he have at the end of 3 years from this investment?

165. Suppose that you invested $13,500 for 9 months earning $450 simple interest. What was the annual rate of interest?

166. If one number is twice as large as another number and the smaller number is increased by 12, the result is 8 less than the larger number. What is the larger number?

167. Katie has received the following scores on her first four math tests: 90%, 83%, 75%, and 76%. What minimum score does she need to receive on her fifth math test in order to have a test average of at least 80%?

168. A bakery sells a box of six blueberry scones for $8.99. If the bakery decides to sell the scones in boxes that hold eight scones, how much should it charge?

169. A company hires a statistician who earns an average starting salary of $45,700 annually, but the actual salary can vary from the average by as much as $2,750. Write an absolute value inequality that describes the salary range. Then solve the absolute value inequality to find the range of starting salaries.

170. Nuts costing $1.40 per pound are mixed with nuts costing $1.00 per pound to produce 40 pounds of mixture worth $1.10 per pound. How much of each type is used?

171. Kyle is filling his bathtub to take a bath. If he turns on only the hot water, the tub will fill in 20 minutes. If he turns on only the cold water, the tub will fill in 10 minutes. How long will it take to fill the tub if Kyle turns on the hot and cold water faucets at the same time?

Answers

Chapter 1

1. $0, 2, \frac{1}{2}, 0.6$
2. 2, 11, 17
3. $\frac{3}{2}$ or $1\frac{1}{2}$
4. False
5. True
6. True
7. False
8. B
9. 3^2
10. 2^8
11. The statement is incorrect because when multiplying powers with the same base, you take the sum of the exponents and keep the same base. So, $7^2 \cdot 7^5$ is equivalent to 7^7.
12. $n = 4$
13. 3^{20}
14. $\frac{4}{9}$
15. B
16. $-\sqrt{64} < -\sqrt[3]{125}$
17. Approximately 4.2
18. Irrational number. The sum of a rational number and an irrational number is irrational.
19. $\sqrt{4}$ was not simplified to 2. The correct solution is $2\sqrt{10}$.
20. 38

21. 2, 3, 4, 5, 6, 8

22. No, both numbers cannot be positive. At least one number needs to be negative for the sum of two numbers to equal a negative number. Three possible numerical expressions are $-2 + -1$, $-5 + 2$, and $6 + -9$.

23. 975 feet below sea level or -975 feet

24. negative

25. $\frac{29}{15}$ or $1\frac{14}{15}$

26. $3\frac{3}{8}$ cups

27. $6.2 \approx 6$, so six chairs can be re-upholstered with $20\frac{2}{3}$ yards of fabric.

28. 5,287 feet

29. 9.94 pints of blood

30. $\frac{4}{9}$

31. 5%

32. $1,594.13

33. $x(1 - 0.13)^3 = x(0.87)^3 \approx 0.66x$

34. 0.423 is not a number in the range [1, 10).

35. $12 \times 10^5 = 1.2 \times 10^6$

36. 4×10^{-6}

Chapter 2

37. $A = \{0, 2, 4, 6, 8\}$ and $B = \{1, 2, 3, 4, 6\}$

38. $A' = \{1, 3, 5, 7, 9, 10\}$

39. $A \cap B = \{2, 4, 6\}$

40. $A \cup B = \{0, 1, 2, 3, 4, 6, 8\}$

41. True

42. $2n - 5$

43. $\frac{1}{2}(n+15)$

44. $\frac{x}{3}+5y$ or $\frac{1}{3}x+5y$

45. The sets have no elements in common.

46. 22

47. The value of h will increase by 30.

48. $38 + 7x$ is the expression for the total cost of entering the amusement park. If the total cost was \$80, six children went to the amusement park.

49. $\frac{27}{100}$ or 27% of a gallon of paint to paint all the cubes for a school play

Chapter 3

50. Addition property of equality is applied first; multiplication property of equality is applied second.

51. D

52. Your friend is incorrect. The equation $-3(2 - x) = 3x + 6$ has no solution since $-6 + 3x = 3x + 6$ and $-6 \neq 6$.

53. $x = \frac{3c}{2-a}$

54. $c = \frac{ad}{b}$

55. $x = -17.5$

56. The equation was solved incorrectly because the 5 was not distributed to the 8 and the like terms, $2x$ and $-3x$, were not combined correctly. The correct solution is $x = -6$.

57. $\ell = \frac{SA - \pi r^2}{\pi r}$ or $\ell = \frac{SA}{\pi r} - r$

58. 2 cm

Chapter 4

59. $x = -1, y = 2$

60. $m = 5, n = 2$

61. Yes. If the graphs of the equations are the same, then there are infinitely many solutions for both equations.

62. No, the point $(1, -2)$ is a solution to $2x + 3y = -4$, but not to $4x - y = 2$.

63. $a = 1$

64. No solution

65. One solution

66. Infinite solutions

67. There is no solution because the lines are parallel and will never intersect.

Chapter 5

68. $-3xy^2$; degree $= 3$

69. $-135x^{14}y^4$

70. $3xyz$

71. $\frac{y}{x^5z^2}$

72. $2a^2 - b - 13$

73. $4x^2y$ and $-3xy^2$ are not like terms because they do not have the same variable factors.

74. $14x + 10y$

75. $5x^2 + 28x + 36$

76. $6mn + 7n$

77. $(x - 6)$ inches

78. $2y(y - 4)$

79. $(3x - 4)(3x + 4)$

80. Two possible coefficients are 12 and 15. With these coefficients, the possible solutions are $x^2 - 12x + 36 = (x - 6)^2$ and $x^2 - 15x + 36 = (x - 12)(x - 3)$. (Answers may vary.)

81. Three possible dimensions of the rectangular prism are $2a$, $(a + 1)$, and $(2a + 1)$. (Answers may vary.)

82. $(m + n)(m + 4n)$

83. $(1 + b)(a + 6)$

Chapter 6

84. $\frac{x^2}{3}$

85. $\frac{x-1}{3}$

86. $\frac{3x^2y}{8}$

87. $2(x + 2); x \neq -1, x \neq -2$

88. $\frac{3xy}{7}$

89. 3

90. The numerator and denominator have no common factors other than 1, so $\frac{x-1}{x+1}$ is simplified.

91. This is not correct because you can factor out -1 in the numerator, so

$$\frac{-1\cancel{(-x+2)}}{\cancel{2-x}} = -1.$$

92. $\frac{2\sqrt{x}-5}{x^2+2x-1}$ is not a rational expression because the exponent in the numerator is not a whole number.

93. $\frac{5y+7x}{xy}$

94. $\dfrac{3a+2b}{a^4b^3}$

95. $\dfrac{x(x+9)}{(x-3)(x+1)}$ or $\dfrac{x^2+9x}{x^2-2x-3}$

96. $\dfrac{5x^2-15x}{(x+4)(x-4)(x+1)}$ or $\dfrac{5x(x-3)}{(x+4)(x-4)(x+1)}$

97. $t=\dfrac{145.8}{r}$

98. $\dfrac{x}{2(x+4)}$

99. The area of the triangle represents $\frac{1}{6}$ the area of the square.

Chapter 7

100. $a = -5$

101. $(-\infty, 6]$

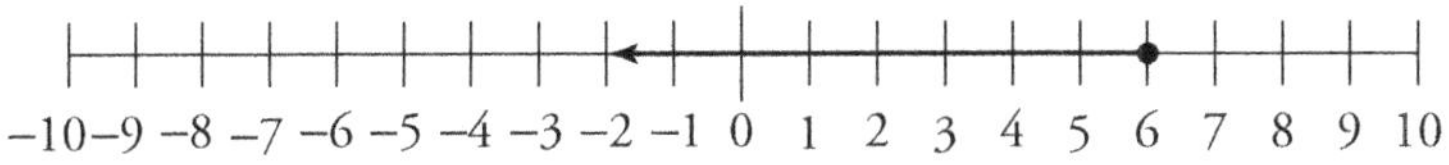

102. 32°F $\le x \le$ 212°F

103. D

104.

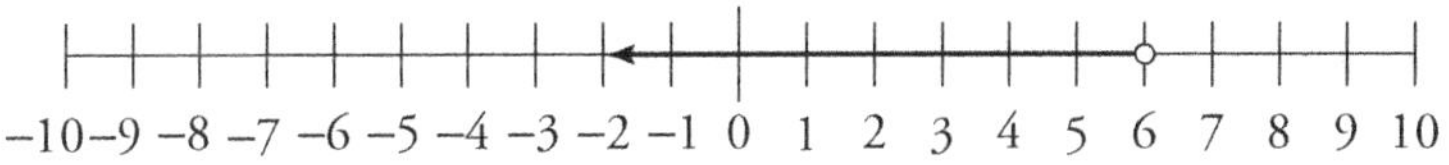

105. 9

106. $|x - 6.5| = 1.5$

107. $x = 6$ or $x = -2$

108. $x > 1$ or $x < -3$

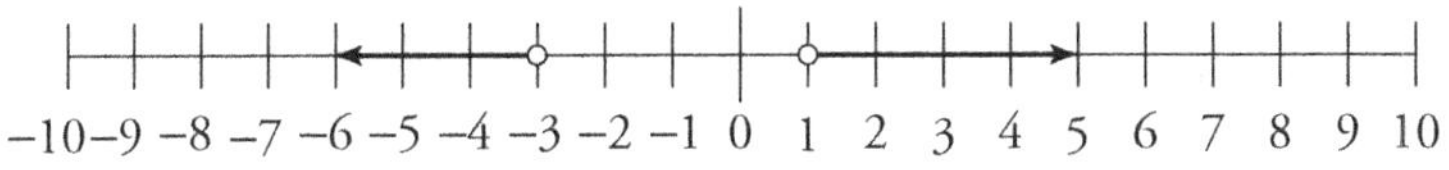

Chapter 8

109. Non-linear, because the equation cannot be written in the form $ax + by = c$

110.

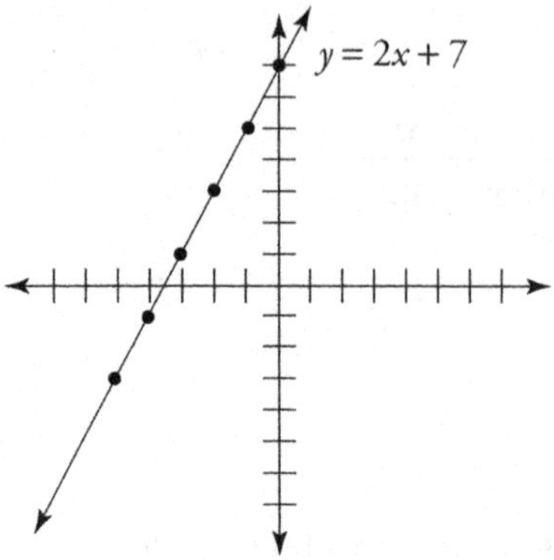

111.

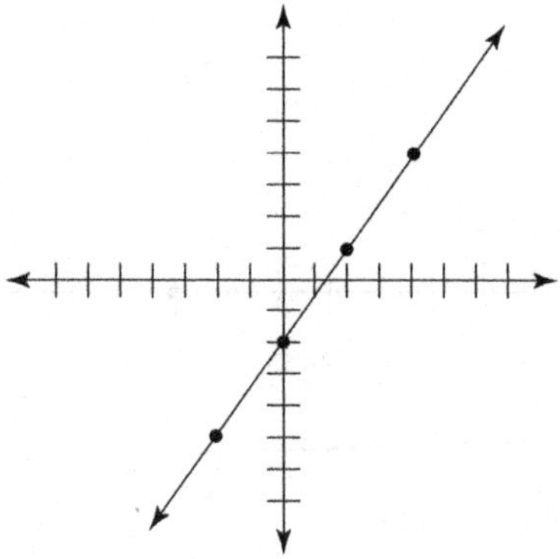

112. $m = -2$

113. $x = -4$

114. Yes

115. D

116. $y = 2x - 7$

117. In point-slope form, the equation is $y+3=\frac{1}{3}(x-6)$, and in standard form, the equation is $x - 3y = 15$.

118. $F = \frac{9}{5}C + 32$

119. $2x - y = 1$; the x-intercept is $\frac{1}{2}$.

120. $G(m) = -0.075m + 20$ or $G(m) = -\frac{3}{40}m + 20$. It will take approximately 4.4 hours to empty the tank. Therefore, the tank will be empty at about 12:26 p.m.

121. The slope, $-5,000$, means that the oil is pumped out at a steady rate of 5,000 barrels per year. The y-intercept, 400,000, represents the initial oil in the oil field, or 400,000 barrels.

122. The relationship is not linear because the rate of change is not constant.

123. $P(x) = 2x + 16$

124.

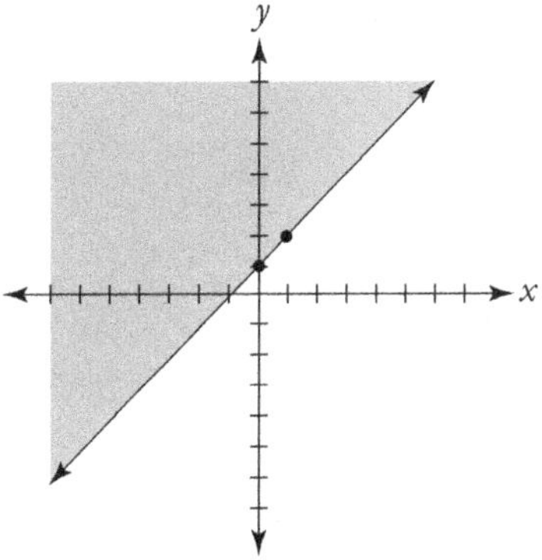

125.

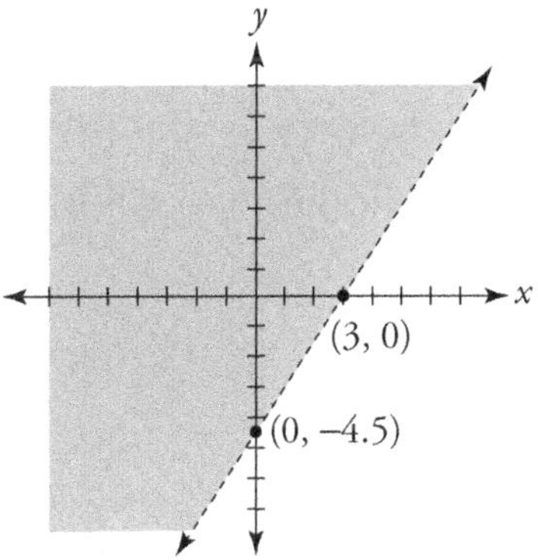

Yes, $(2, -1)$ is a solution to the inequality since $8 < 9$ when you substitute the point into the linear inequality $3x - 2y < 9$.

Chapter 9

126. Graph (a)

127. Since there are two different outputs for the same input, $(-1, 5)$ and $(-1, 0)$, the relation is not a function.

128. 20

129. Range = {8, 4, 0}

130. B

131. B

132. $x = 4$

133. $k = \frac{1}{3}$; $y = \frac{1}{3}x$ or $y = \frac{x}{3}$

134. $y = \frac{1}{4}$

135. $y = \frac{36}{x}$

136. Inverse variation; $k = 28$ and $y = \frac{28}{x}$

137. Direct variation; $k = -4$ and $y = -4x$

138. Indirect variation; the cost each person pays multiplied by the number of people will equal a constant product of \$79.95 (the cost of five pizzas at \$15.99 each).

139. Direct variation; the ratio of the total cost of the oranges in cents and the number of pounds of oranges being bought is constant at 0.99 to 1.

140. Brian should be about 6.1 feet from the fulcrum to balance the seesaw.

Chapter 10

141. $5\sqrt{3}$

142. C

143. $\sqrt{5}$ and $\sqrt{75} = 5\sqrt{3}$ are not like radicals because they do not contain the same radicand.

144. $\dfrac{4}{2-\sqrt{3}}$ is not in simplest form because the denominator contains a radical. The simplified form is $8+4\sqrt{3}$.

145. $5x^2y^4\sqrt{x}$

146. $8\sqrt{2}$

147. $72t^2\sqrt{t}$

148. $13-4\sqrt{3}$

149. $\dfrac{\sqrt{14}}{2}$

150. $\dfrac{8+5\sqrt{2}}{2}$ or $4+\dfrac{5\sqrt{2}}{2}$

151. $w\sqrt{17}$

Chapter 11

152. $x = 9$ or $x = -7$

153. $x = 9$ or $x = -9$

154. $x = 0$ or $x = -8$

155. $x = -1$

156. $x = \dfrac{-3+\sqrt{5}}{2}$ or $x = \dfrac{-3-\sqrt{5}}{2}$

157. $x = \dfrac{-7+\sqrt{33}}{2}$ or $x = \dfrac{-7-\sqrt{33}}{2}$

158. $b = 6$

159. Two possible factors of the quadratic equation are $(2x - 5)$ and $(x - 2)$. The quadratic equation in standard form is written $2x^2 - 9x + 10 = 0$. (Answers may vary.)

160. The discriminant equals 0, which means that there is one real number solution.

161. $t = 2.96$ seconds

162. $(2x + 4)(x + 2) = 130$; the new dimensions of the rectangular garden are 8.1 feet and 16.2 feet.

163. Use the quadratic formula or completing the square because $x^2 + 5x - 2 = 0$ is not factorable.

Chapter 12

164. \$268.98

165. $r = 4.\overline{4}\%$

166. 40

167. Katie needs to earn at least a 76% on her fifth test in order to have a test average of at least 80%.

168. \$11.99 for a box of eight blueberry scones

169. $|x - 45{,}700| \leq 2{,}750$; the range of starting salaries, x, is $\$42{,}950 \leq x \leq \$48{,}450$.

170. 10 pounds of \$1.40 nuts and 30 pounds of \$1.00 nuts

171. $6\frac{2}{3}$ minutes

GLOSSARY

abscissa (*x*-coordinate): The distance along the horizontal axis in a coordinate graph.

absolute value: A number's distance from zero on a number line (the numerical value when direction or sign is not considered). The symbol for absolute value is | |.

addition property of equality: Adding the same number to each side of an equation gives an equivalent equation (if $a = b$ then $a + c = b + c$).

addition property of inequality: Adding the same number to each side of an inequality gives an equivalent inequality (if $a > b$, then $a + c > b + c$).

additive inverse: The opposite (negative) of a number. A number and its additive inverse have a sum of 0.

algebra: A branch of mathematics that uses variables and/or symbols in place of numbers to express mathematical operations and relationships.

algebraic expressions: Mathematical phrases that contain one or more variables.

algebraic fractions: Fractions using a variable in the numerator and/or denominator.

ascending order: When the degree of a polynomial's terms increases for each succeeding term.

associative property: Grouping of elements does not make any difference in the outcome. Only true for multiplication and addition.

binomial: A polynomial consisting of exactly two monomials.

braces: Grouping symbols used after the use of brackets. Also used to represent a set. Braces look like this: { }.

brackets: Grouping symbols used after the use of parentheses. Brackets look like this: [ ].

canceling (cross-cancel): In multiplication of fractions, dividing the same common factor into both a numerator and a denominator.

Cartesian coordinates: A system of assigning ordered number pairs to points on a plane.

closed half-plane: A half-plane that includes the boundary line and is graphed using a solid line and shading.

closed interval: An interval that includes both endpoints or fixed boundaries.

closed ray: A ray that includes its endpoint (closed half-line).

closed sentence (closed statement): An equation that contains numerical expressions. The equation can be either true or false.

closure property: When results of an operation performed on a set of numbers fall into the original set.

coefficient: A number in front of a term that contains a variable(s). For example, in $9x$, 9 is the numerical coefficient of x and x is the variable coefficient of 9.

common factors: Factors that are the same for two or more numbers.

commutative property: Order of elements does not make any difference in the outcome. Only true for multiplication and addition.

complement of a set: All of the elements in the universal set that are not in the set being considered.

complex fraction: A fraction having a fraction or fractions in the numerator and/or denominator.

composite number: A natural number divisible by more than just 1 and itself (such as 4, 6, 8, 9, . . .). Zero and 1 are not composite numbers.

compound inequality: A mathematical sentence with two (or more) inequalities joined by the word "and" or the word "or."

conjugate: The conjugate of a binomial contains the same terms, but the opposite sign between them. $(x + y)$ and $(x - y)$ are conjugates.

coordinate axes: Two perpendicular number lines used in a plane that intersect at the origin (0, 0).

coordinate plane: Two perpendicular number lines, the x-axis and the y-axis, creating a plane on which each point is assigned a pair of numbers.

coordinates: The numbers that correspond to a point on a coordinate plane.

cross products property: In a proportion, where a, b, c, and d are real numbers, $b \neq 0$, and $d \neq 0$, if $\frac{a}{b} = \frac{c}{d}$, then $ad = bc$.

cube number (perfect cube): The result when a number is multiplied by itself twice. Designated by the exponent 3 (such as x^3).

cube root: The number that when multiplied by itself twice gives you the original number. For example, 5 is the cube root of 125: $\sqrt[3]{125} = 5$.

degree of a monomial: The sum of the exponents of its variables.

degree of a polynomial: The highest degree of its terms in the simplified expression.

denominator: The expression below the fraction bar in a fraction.

descending order: When the degree of a polynomial's terms decreases for each succeeding term.

direct variation: When y varies directly as x or y is directly proportional to x.

discriminant: The value under the radical sign in the quadratic formula. $(b^2 - 4ac)$

disjoint sets: Sets that have no elements in common (the intersection is the empty set).

distributive property: The process of multiplying the number on the outside of the parentheses by each term inside the parentheses. $a(b + c) = ab + ac$

division property of equality: Dividing by the same number on each side of an equation gives an equivalent equation. (If $a = b$ and $c \neq 0$, then $\frac{a}{c} = \frac{b}{c}$.)

domain: The set of all x values from the ordered pairs in a relation.

element: A member of a set.

elimination method: A method of solving a system of equations by adding or subtracting the equations to eliminate one of the variables.

empty set (null set): A set with no members (a null set). The symbol for an empty set is $\varnothing$.

equal sets: Sets that have exactly the same members.

equation: A mathematical statement or relationship between numbers and/or variables in which two expressions are of equal value.

equivalent equations: Equations that have the same solution.

equivalent sets: Sets that have the same number of members.

evaluate: To determine the value or numerical amount.

even number: Integer divisible by 2.

excluded values: Values of one or more variables that will make the denominator of an expression equal to zero.

exponent: A number placed above and to the right of a quantity; used to indicate repeated multiplication.

extremes: Outer terms of a proportion. In the proportion $\frac{a}{b} = \frac{c}{d}$, a and d are the extremes.

factor: To find two or more quantities whose product equals the original quantity.

finite: Countable; having a definite ending.

finite set: Contains a countable number of elements.

F.O.I.L. method: A method of multiplying binomials in which first terms, outside terms, inside terms, and last terms are multiplied, in that order.

fraction: A number in the form $\frac{a}{b}$, where a represents an integer and b represents an integer other than zero.

function: A relation in which each element in the domain is paired with exactly one element in the range.

graphing method: A method of solving a system of equations by graphing each equation on a coordinate graph and finding the point or points of intersection.

half-open interval: An interval that includes one endpoint, or one boundary.

half-plane: The region of a coordinate plane on one side of a boundary line.

horizontal line: Equation of a line in the form $y = k$, where k is a constant.

identity equation: An equation that is true for all values of the variable(s); the solution set is the set of all real numbers.

identity element for addition: The identity element for addition is 0. The sum of any number and 0 gives the original number. Also called *additive identity*.

identity element for multiplication: The identity element for multiplication is 1. Any number multiplied by 1 gives the original number. Also called *multiplicative identity*.

imaginary numbers: Square roots of negative numbers. The imaginary unit is i.

incomplete quadratic equation: A quadratic equation with a term missing.

inverse variation (indirect variation): When y varies indirectly with x or y is indirectly proportional to x. That is, as x increases, y decreases and as y increases, x decreases. Also referred to as inverse or indirect proportion.

inequality: A statement in which two mathematics expressions are not equal.

infinite: Uncountable; continues forever.

infinite set: Contains an uncountable number of elements.

integer: A number that can be positive, negative, or zero without any decimal or fractional parts.

intersection of sets (two or more): A set containing only the elements that are in both sets (overlap in a Venn Diagram).

interval: All the values that lie within two certain boundaries.

interval notation: Uses an interval to represent a pair of boundary numbers with brackets and/or parentheses.

inverse operations: Opposite operations that undo each other (e.g., addition and subtraction, multiplication and division).

inverse relations: Relations where the domain and the range have been interchanged—switching the coordinates in each ordered pair.

irrational number: Cannot be written as a ratio of two integers, $\frac{a}{b}$, where $b \neq 0$. The decimal equivalent of the number neither terminates nor has a repeating decimal pattern.

like terms: Monomials that have the same variable factors.

linear equation: An equation in two variables whose solution set forms a straight line when plotted on a coordinate plane.

literal equation: An equation that contains two or more variables.

means: Inner terms of a proportion. In the proportion, $\frac{a}{b} = \frac{c}{d}$, b and c are the means.

monomial: An algebraic expression consisting of only one term. This term can be a real number, a variable, or a product of real numbers and variables with whole-number exponents.

multiplicative inverse: The reciprocal of a number. Any nonzero number multiplied by its multiplicative inverse equals 1.

multiplication property of equality: Multiplying by the same number on each side of the equation gives an equivalent equation (if $a = b$, then $ac = bc$).

natural number: 1, 2, 3, 4, and so on. (Also referred to as a counting number.)

negative division property of inequality: Reverse the inequality sign when dividing by a negative number (if $a > b$ and $c < 0$, then $\frac{a}{c} < \frac{b}{c}$).

negative multiplication property of inequality: Reverse the inequality sign when multiplying by a negative number (if $a > b$ and $c < 0$, then $ac < bc$).

non-ending interval: An interval where the values to one side of an endpoint extend on forever.

non-linear equation: An equation whose solution set does not form a straight line when plotted on a coordinate plane.

null set: A set with no members (an empty set). The symbol for the null or empty set is $\varnothing$.

number line: A graphic representation of real numbers. The point on this line associated with each number is called the graph of the number.

numerator: The expression above the fraction bar in a fraction.

numerical coefficient: A number in front of a variable term in an algebraic expression.

odd number: Integer that is not divisible by 2.

open half-plane: A half-plane that does not include the boundary line. If the inequality is a "$>$" or "$<$", then the graph is an open half-plane.

open interval: An interval that does not include endpoints or fixed boundaries.

open ray: A ray that does include its endpoint (open half-line).

open sentence: An equation that contains one or more variables. The equation is neither true nor false and the solution is unknown.

ordered pair: Any pair of elements (x, y) having a first element x and a second element y. Used to identify or plot points on a coordinate grid.

ordinate (*y*-coordinate): The distance along the vertical axis on a coordinate graph.

origin: The point of intersection of the two number lines on a coordinate graph. Represented by the coordinates (0, 0).

percent: A ratio that represents parts per one-hundred.

point-slope form of a linear equation: The equation of a line in the form $y - y_1 = m(x - x_1)$, where m is the slope of the line and (x_1, y_1) is any point on the line.

polynomial: An algebraic expression consisting of one or more monomials.

positive division property of inequality: If $a > b$ and $c > 0$, then $\frac{a}{c} > \frac{b}{c}$.

positive multiplication property of inequality: If $a > b$ and $c > 0$, then $ac > bc$.

prime number: A natural number greater than 1 that has only two factors, 1 and itself (such as 2, 3, 5, 7, . . .).

properties of equality: Basic rules for using the equal sign.

proportion: Equation written as two equivalent ratios in the form $\frac{a}{b} = \frac{c}{d}$, where $b \neq 0$ and $d \neq 0$. For example, a is to b as c is to d.

quadrants: Four quarters or divisions of a coordinate plane.

quadratic equation: An equation that could be written in standard form as $ax^2 + bx + c = 0$, where a, b, and c are real numbers and $a \neq 0$.

quadratic formula: A method of solving quadratic equations using the formula $x = \frac{-b \pm \sqrt{b^2 - 4ac}}{2a}$.

radical sign: The symbol used to designate square root: $\sqrt{\ }$ (or a higher root when superscript is inserted in front of the radical sign).

range of a relation: The set of all y values from the ordered pairs in a relation.

ratio: A method of comparing two or more quantities by division. For example, $a{:}b$. Often written as a fraction, $\frac{a}{b}$.

rational expressions: Fractions in which the numerator and/or the denominator are polynomials.

rational number: A positive or negative number that can be written as a ratio of two integers, $\frac{a}{b}$, where $b \neq 0$. These numbers include integers and terminating and repeating decimals.

real numbers: The set consisting of all rational and irrational numbers.

reciprocal: The reciprocal of a number is one divided by the number. Also referred to as multiplicative inverse.

reducing (simplifying): Changing a numerical or algebraic fraction into its lowest terms. For example, $\frac{2}{4}$ is reduced to $\frac{1}{2}$, or $\frac{a}{ab}$ is reduced to $\frac{1}{b}$.

reflexive property of equality: Any number equals itself ($a = a$).

relation: Any set of ordered pairs.

repeating decimal: A decimal corresponding to a fraction that continues forever repeating a number or block of numbers.

roster form: A method of naming a set by listing its members.

rule: A method of naming a set by describing its elements.

scientific notation: A method of writing a very large or a very small quantity in terms of a decimal number greater than or equal to 1 or less than 10, multiplied by a power of 10.

set: A group or collection of distinct objects, numbers, and so forth.

set-builder notation: A formal method of describing a set. Often used for inequalities. For example, $\{x| \, x > 1\}$, which is read "x such that all x is greater than 1."

simplify: To combine several or many terms into fewer terms.

simultaneous equations (system of equations): A set of equations with the same unknowns (variables).

slope-intercept form of a linear equation: The equation of a line in the form $y = mx + b$, where m is the slope of the line and b is the y-intercept.

slope of a line: The ratio of the change in y to the change in x in a linear equation (slope = $\frac{\text{rise}}{\text{run}}$).

solution set (solution): All the answers that satisfy the equation.

square number (perfect square): Nonzero integer that results when a number is multiplied by itself. Designated by the exponent 2 (such as x^2).

square root: The number that when multiplied by itself gives you the original number. For example, 5 is the square root of 25: $\sqrt{25} = 5$.

standard form of a linear equation: The equation of a line written in the form $Ax + By = C$, where, if possible, A, B, and C are integers and A and B are not both zero.

subset: A set contained within another set.

substitution method: A method of solving a system of equations that involves substituting one equation into another.

subtraction property of equality: Subtracting the same number from each side of an equation gives an equivalent equation (if $a = b$, then $a - c = b - c$).

subtraction property of inequality: Subtracting the same number from each side of an inequality gives an equivalent inequality (if $a > b$, then $a - c > b - c$).

symmetric property of equality: If the first number is equal to the second number, then the second number is equal to the first number (if $a = b$, then $b = a$).

system of equations (system): A collection of two or more equations with the same set of variables.

term: A number, variable, or a product of numbers and variables.

transitive axiom of inequality: If $a > b$ and $b > c$, then $a > c$. Or if $a < b$ and $b < c$, then $a < c$.

transitive property of equality: If the first number is equal to the second, and the second is equal to the third, then the first number is equal to the third (if $a = b$ and $b = c$, then $a = c$).

trichotomy axiom of inequality: The only possible relationships between two numbers are $a > b$, $a = b$, or $a < b$.

trinomial: A polynomial consisting of exactly three monomials.

union of sets: A set containing all the elements in two or more sets.

universal set: The general category set, or the set of all those elements under consideration.

unknown: A letter or symbol whose value is not known; a variable.

value: Numerical amount.

variable: A symbol, often a letter, used to represent an unknown number.

variation: A relation that shows how one quantity changes relative to another quantity.

Venn diagram: A pictorial description that uses circles or ovals to represent two or more sets and their relationships.

vertical line: Equation of a line in the form $x = h$, where h is a constant.

vinculum: A line placed over (sometimes under) a digit or group of digits in a repeating decimal fraction to show which digits are repeating. For example, $\frac{1}{3} = 0.\overline{3}$.

whole number: 0, 1, 2, 3, and so on.

***x*-axis:** The horizontal axis in a coordinate plane.

***x*-coordinate (abscissa):** The first number in the ordered pair. Refers to the distance from 0 on the x-axis.

***x*-intercept:** The point where a graph on the coordinate plane crosses the x-axis.

***y*-axis:** The vertical axis in a coordinate plane.

***y*-coordinate (ordinate):** The second number in the ordered pair. Refers to the distance from 0 on the y-axis.

***y*-intercept:** The point where a graph on the coordinate plane crosses the y-axis.

zero-product property: If a and b are real numbers and $ab = 0$, then $a = 0$ or $b = 0$, or both $a = 0$ and $b = 0$.

Made in the USA
Monee, IL
04 November 2026

67384046R00154